CHANCE IN THE
HOUSE OF FATE

CHANCE IN THE HOUSE OF FATE

A NATURAL HISTORY OF HEREDITY

JENNIFER ACKERMAN

BLOOMSBURY

First published in Great Britain 2001
This paperback edition published 2002

Copyright © 2001 by Jennifer Ackerman

The moral right of the author has been asserted

Grateful acknowledgment is made for permission to reprint material from
The Complete Essays of Montaigne, translated by Donald M. Frame,
published by Stanford University Press. Copyright © 1958 by the Board of
Trustees of the Leland Stanford Junior University, used with the
permission of the publishers.

Bloomsbury Publishing Plc, 38 Soho Square, London W1D 3HB

A CIP catalogue record for this book
is available from the British Library

ISBN 0 7475 5820 5

10 9 8 7 6 5 4 3 2 1

Printed in Great Britain by Clays Ltd, St Ives plc

FOR MY FAMILY

CONTENTS

Part IV
PASSAGE

I am the family face;
Flesh perishes, I live on.
— Thomas Hardy

PREFACE

THERE ARE MYSTERIES in all families. Those that arrest me, that set me back on my heels, are the mysteries of heredity—the past whispered in bone and blood; the dozens of ancestors rolled up in one skin, to be read in "curve and voice and eye," as Thomas Hardy wrote, "the seeds of being that heed no call to die" but turn up again and again on the doorstep like a ne'er-do-well uncle. It seems astonishing that a sweep of eleven generations hardly modifies the night blindness of one family or the trembling jaw of another, that fifty or a hundred years may fail to alter a familial pattern of whorled eyebrow or "wolf's" teeth, the musical genius of the Bach family, or the dimpled chin of my husband's tribe.

In the last decade or so, a startling new message has come out about the long hold of heredity. Members of the human family carry traits that have held on down the line not just for generations but for eons, traits that mock all boundaries of time and kind. Scientists probing the deep workings of organisms from yeast to humans have turned up news that despite our outward differences of life and limb, we are run by similar genes and proteins, similar cell parts and mechanisms, which have weathered evolution over ages, passing nearly intact through hundreds of millions of years of rising and falling forms. These shared molecules and routines affect nearly all the turnings of life, from birth and growth to perception and behavior.

This book is a pilgrimage to the heart of heredity. It is a natural history not in the literal sense of a systematic inquiry, but rather in the etymological sense, a telling of stories about life, lineage, chance, and fate; about family, kin, and kind. It explores both the projecting traits of the human family—the one we're born into and the one we create—and also the bigger, deeper inheritance that ties us to the rest of life in profound, even shocking ways.

. . .

I like to hang around the doorway of biological surprise. For years I have collected news of curious findings, of young spiders that eat their mothers, of a giant fungus infecting miles of Michigan forest spawned by a single spore in the last ice age, of fish with fingers, caterpillars with lungs, genes with secrets. I don't profess to worship everything, but I do harbor strange sympathies fired by such discoveries, a kind of naturalist's faith. This is the news that sweeps me away, the gnomic workings of the living order, nature's inventive jack-in-the-box surprises that shift our view of life like the sudden twist of a kaleidoscope.

Here is an item from my files. When scientists deciphered the intimate details of mating in yeast, that single-celled fungus that raises our bread and brews our beer, they got a shock. The molecule that draws two yeast cells into sex closely resembles one made by our own brain cells to regulate reproduction.

The likeness seemed a fluke at first. But then other examples popped out of the box: genes that shape the bodies of fruit flies so like our own body-shaping Hox genes that one can put a human Hox gene into a developing fruit fly embryo, and it will carry out the job of the fly's gene without a hitch; genes that shape the marvelous globe of the human eye strangely similar to those that carve the compound eye of a fruit fly; the tiny genetic mechanisms that drive our biological rhythms, keeping us in tune with the big swings of night and day, matching those in algae. So, too, do we share with other organisms the ancient genes that dictate cell death, the phenomenon that underlies metamorphosis, turning tadpoles into frogs and caterpillars into butterflies and also shapes our bodies, whittling away the webbing between fingers before birth, eliminating inappropriate sexual organs. Common to all of us, as well, is a suite of small, sturdy messenger molecules, offering clues to such mysteries as why the cells of the human brain respond to the chemical messages of the poppy plant and to the potent sexual attractants of a Himalayan deer.

What are chemicals found in the human body doing in plants, fungi, bacteria? How can genes that shape a fruit fly be near twins of my own?

Disparate organisms, it seems, are more radically alike than we ever imagined. Our deepest selves—our very cells and molecules—are alive with reminders of old, enduring connections with other creatures, resemblances that run right down to the root of the tree of life. These items of shared inheritance have formed a library of wonders in my mind's eye. That there is a certain sameness among life's various forms follows from the notion that we all arose, ultimately, from a common ancestor. We are shaped by fate, by what came before. But life has chanced to venture in wildly different directions. In learning to suck energy from sunlight and in swallowing shocking amounts of oxygen, in heaving up from the beneficent chemical crucible of the sea and in exploring leafy interiors and desiccated desert, life has split into discrete identities, strewn about fresh designs, unimaginably varied feet, teeth, tongues, antennae, wings, leaves, brains.

In this world of dreamlike change, the lexicon of genes, like human languages, is thought to evolve along unreturning tracks. We know that nature is constantly making random changes in almost all genes, and that two species that diverged from a common ancestor hundreds of millions of years ago are likely to have accumulated a lot of little alterations. As eons pass, so do variants of genes, vanishing on the same wind that took the tyrannosaurs. It seems strange and wonderful that among organisms so spirited with individuality and detail—pepper frog, salp, dragonfish, basset hound—there should be so much solid common ground.

Over the last few years I have wandered the body, looking for these legacies and slim continuances, seeking to ferret them out of their holes and sun them a little, to brush their surface in places, give them a stab or pinch them to the bone if I could. I have tracked the labyrinthine world of laboratories, too, asking about the molecular bricks that underlie the splendid medley of living forms: what makes them work so beautifully that they have demanded little change in hundreds of millions of years? What happens if they go awry? If organisms of such diverse stripes are made of similar genes, how is newness born in the world?

By exploring this deep-down world, I hope to create new shelves in my mind for the recent profusion of genetic discoveries, the news of the sequencing of genomes from the tubercle bacillus to *Homo sapiens,* the findings of genes linked with cancer, Alzheimer's disease, migraine, and baldness, passed down from father to son, grandmother to granddaughter; genes affecting intelligence, sexual preference, spatial ability, anxiety, sense of well-being—some of them discovered in small, so-called model organisms such as worms, fruit flies, mice.

What is a gene, anyway? Are there genes "for" particular traits? Are the letters DNA and RNA an Open Sesame to all the familial secrets of life? Can we starve all of nature's mysteries into molecular oneness, explain the fruit solely by its root?

And what does one make of the notion that our genes mirror those in yeast? Two decades ago scientists discovered that humans and chimpanzees appear to have in common about 98 percent of their DNA. Chimps are one thing, yeast is quite another. The news that, when it comes to molecules, we are so perilously close to our tailed, finned, and spoorish brethren goes against the stories I grew up on, biblical tales of human supremacy and uniqueness, stories of how I was "fearfully *and* wonderfully made," as it is written in the Psalms, to get up before the sun and buy a river, to buzz above all creatures, "over the fish of the sea, over the fowl of the air, over the cattle, over every creeping thing that creepeth upon the earth."

Fear not therefore: ye are of more value than many sparrows.

(Luke 12:7)

We have for so long picked ourselves out from the horde of other creatures, reckoned ourselves the peak and point of nature's whole history. What to do now, with this news of our deep-down similarity, our profound kinship, with "lowly" organisms?

The physicist Michio Kaku once wrote that finding the key to weather and seasons required a leap into another dimension, up into outer space. Understanding humanity's place in the matrix of life requires just such a leap, but downward, into the diminutive world of genes and cells.

Raised as I was on gerbils and birds, on the love of the whole organism, not its microscopic parts, I find it a stretch to descend into the darkness of a molecular world. I know my bats, weasels, and wood frogs far better than I do the crabbed atoms of a hemoglobin molecule. I am far more comfortable exploring the elements of the violet family than those of the periodic table. The human mind may have mastered the black hole and the quark, but most of us have difficulty grasping the very big and the very small. We tend to think easily only of things on our own scale, midway between the atom and the sun. The first microscopists, confronted with the bizarre creatures swimming beneath their lenses, sought desperately to see bodies like their own, searched for sign of head or tail, denied as long as they could the many orifices and multiple stomachs, the brainless chunks of transparent flesh. So, too, we may seek in vain the familiar in the minute parallel planet of genes and proteins.

To make things worse, the language of this world veers into the cold domain of chemistry, where the common nouns are "nucleic acids" and "amino acids"; the common verbs, "regulate," "synthesize," and "catalyze." One scientist grappling with the absence of a precise definition for the term "gene" offered this to snarl the brain: "It is the nucleotide sequence that stores the information which specifies the order of the monomers in a final functional polypeptide or RNA molecule, or set of closely related isoforms."

But despite the dull terminology used to describe it, the cosmos of molecules and cells has surprising beauties and minute dramas every bit as beguiling as those of a bushmaster or a Bengal tiger. In DNA, proteins, even in the molecules of water encapsulated in our cells, are shapely details, beautiful clues that hold the key to everything from the acuity of the eye to the memory of the immune system. In their daily workings are tales of seduction, compromise, duplicity, deception, stubbornness, art, magic, death.

I first learned of the Hox body-shaping genes when I was a few months pregnant with my second child. The idea that the molecular mechanisms shaping my baby's growth were the same as those fashioning the fruit fly I found oddly comforting. Think of all the bending and breaking in the boughs of life. The notion that species as remotely related as humans and flies are shaped by the same genes — genes that have slipped in and out of the Cambrian, the Devonian, the Permian, the Pleistocene, requiring little revision in all that time — suggests that they must perform their task beautifully and will not easily be wrenched off course.

Fish, fruit flies, wondrous babies: we may be a feast of distinct entities, but we share the odd economies of nature from birth to death. I'm thrilled to find that we're connected with other organisms, not by something as vague or slippery as animal nature, but by a strong ribbon of measurable molecules, molecules so alike that they can be swapped between species separated by half a billion years of evolution.

I think our minds are built for the pleasure of discovering likenesses or links between vastly different things. It is why we delight in learning that the words "fate" and "symphony" share an ancient root meaning "to speak"; that the opening of Beethoven's Fifth neatly repeats the call of the white-breasted wood wren; that pointing a single finger to draw attention to something of interest is bound tightly to the learning of language. (The earlier a baby extends a demonstrative digit, the more words he or she will know by the age of two.) It is why we love syzygies and rhymes and why we are undone by Romeo's words when he finds Juliet in the tomb and thinks her dead: "Death hath suckt the honey of thy breath." It is why we believe Emerson when he tells us that "the world is a Dancer; it is a Rosary; it is a Torrent; it is a Boat; a Mist; a Spider's snare."

The language of science holds a hunch here. Though some scientific terms are Latinate and pompous, or simply weedy (*deoxyribonucleic acid,* for instance, a great millipede of a term that puts the mind off with its literalness), there are other terms — pithy, germinal, long-lived, and prophetic — that link the unlike and suggest the blooming mysteries of both language and life. The word "gene" goes back to an Indo-European root word that meant beginning and birth. This gave rise to the Old English *gecynd,* meaning family, kin, or kind. The Greek and Latin variants blossomed into a bunch of *gen* words with a multitude of jobs: genus, genius, gender, gentle, generous, generation, genealogy, genesis. One Latin stem became *gnatus,* unfurling into innate, native, natural.

That so short and spare a word as "gene" would persist through the revolutions of language and pop up in all these new, masterful forms impresses me. So do these shared ancestral genes, which are something like word roots. Knowing them is a way of prizing what is essential in our common heritage. That we are still abob with these ancient bits of biological wisdom, that they have endured over eons in creatures as genealogically distant as worms and widowed aunts, is to me as much a cause for celebration as a Bach cantata or bird song.

These fragments of shared biology arose by chance and became fate. I have come to think of them as points of entry or small portholes through which to view the natural history of heredity. Or, perhaps, like the scriptural mustard seed cast into the family garden, from which one might draw radii to every corner of nature.

This book is a tracing of those radii. It is a journey in four parts, starting with the roots of all flesh, the inherited molecules that keep our bodies and those of other mortals alive and thriving; then moving to the generation of our being as single, sentient organisms, from our beginnings in sperm and egg — those seeds of inheritance — to the birth of bodies with vision and the capability of sex. Thence to our relations with other living things, how we recognize, compete, and conspire with them in the deepest, most intimate ways to make something more useful, more skillful, more beautiful than what we might have made alone. And finally to our common passage through time, from the immediate tick of the present, by which our bodies stay in tune with the swing of sun and moon, to our long passage out of the past, from earliest beginnings.

PART I

ROOTS

Root, from Indo-European *werād*, branch or root. The inner core or essential nature of something; a source or origin; an antecedent or ancestor.

1

GENEALOGY

WHY IS IT so strange and sweet to ponder a family tree? There's the chance to peer beyond one's personal limits, to spot from the mast a sea of unremembered relatives. Or to hunt up a lost connection with a distant great-grandmother, to delve into the mystery of her existence in her old home country. There's the hope for pride of birth or social credentials or a sense of rootedness, an antidote for those plagued by mobility, by birdly migrations over one ocean or another, between country and town, up or down the social ladder. Then there's the slim but tantalizing possibility of pecuniary gain in uncovering a link with a relative of large estate — though this is often a mixed blessing, *damnosa hereditas,* the Roman jurist Gaius called it. Come for your inheritance, says a Yiddish proverb, and you may have to pay for the funeral.

In my own family tree, I've relished the weird weaving of ancestral names — Doerfler, Dresen, Homann, Huck, Koeppel, on one side; Goldfarb, Dunkelmeyer, Blank, on the other — names of dimly recollected forebears who watched the skyline of New York from the fourth-class deck of a steamer. I've also applauded the surprises in my immediate family, the four girls defying the odds of sex ratio and, especially, the exotic tendril of our fifth sister, Kim, adopted by my parents from a Korean orphanage when she was four, whose own biological roots remain a kind of Siberia.

Apart from satisfying a curiosity about one's origins or a yearning for old connections, apart from settling the matter of estate, a practical use of genealogy is to sort out the hereditary components of disease. When I was entering puberty, my mother took my sisters and me to a genetic counselor to discuss our risk of bearing defective babies. Ten years earlier my youngest sister, Beckie, had come into the world with microcephaly and profound mental retardation.

Beckie had my mother's slim build and dark hair, my father's brilliant blue eyes, and a brain so stunted that it would never allow her to progress beyond a developmental age of six months. Beckie bore witness to the curious dispensation of nature and gave me a kind of fierce pride in differences. She made us love her from the start, with her innocence wide open like water or air, her deep fondness for music and peekaboo games, her sweet, easy affection and capacity for passionate attachment, especially to my mother.

But imagine expecting another wonder baby and getting one that couldn't turn over, couldn't find a nipple, who later couldn't crawl, babble, point an inquiring digit. When a lioness gives birth to a defective cub, she slams it against the ground until it's dead. The experts advised institutionalizing Beckie. My mother, who had recently pulled a child — my sister Kim — out of one institution, could not accept the idea of putting this child into another. Generous to the bone, she could not but treat Beckie tenderly, carefully, equally, as her own, with honor.

Our extended family — grandmothers, cousins, uncles — felt as Gregor Samsa's family did in Kafka's *Metamorphosis*. When one morning Gregor was transformed into a cockroach, his family reassured itself again and again that "things will be fine, that he will come back to us." Surely our family could find the quick fix, the turning of some deep chemical key that would bring Beckie back to normal, the swift knock to the special lobe or zone in her skull that would cause a little cerebral explosion of nova-like growth to set things right.

We knew better. Beckie was the hatchling of our new reality. To some of us she gave a bright, burning belief in the sanctity of life; to others, a smoldering point of shame or resentment, stoked by the deep fear that we, too, might one day bear such a changeling.

Doctors had ruled out as possible causes of Beckie's deformities alcohol, drugs, exposure to lead, Down's syndrome, brain damage from insufficient oxygen or cerebral bleeding during birth. My mother now hoped to rule out any play of heredity.

In the sixteenth century, Michel de Montaigne counted heredity "among the wonders so incomprehensible that they surpassed even miracles in obscurity."

What a prodigy it is that the drop of seed from which we are produced bears in itself the impressions not only of the bodily form but of the thoughts and inclinations of our fathers. Where does that drop of fluid lodge this infinite number of forms? And how do they convey these resemblances with so heedless and irregular a course that the great-grandson will correspond to his great-grandfather, the nephew to the uncle?

In one Roman family, he observed, "there were three, not in a row but at intervals, who were born with the same eye covered with cartilage. At Thebes there was a family that from their mother's womb bore the figure of a lance-head, and whoever did not bear it was considered illegitimate." Montaigne believed that he himself had inherited from his father that "stony propensity," the painful curse of gallstones. He puzzled over how his father, who had first begun to suffer from gallstones twenty-five years after his son was born, could have transmitted to his offspring something he did not possess when the son was conceived.

Where was the propensity to this infirmity hatching all this time? And . . . how did this slight bit of his substance, with which he made me, bear so great an impression of it for its share? And moreover, how did it remain so concealed that I began to feel it forty-five years later, the only one to this hour out of so many brothers and sisters, and all of the same mother?

In preparation for the genetic counseling, I puzzled over charts of traits dominant and recessive, diagrams of red- and white-eyed flies and of the smooth and dimpled peas that illustrated Gregor Mendel's brilliant insights into genetic dominance and the segregation of traits. Before Mendel, heredity was thought to be transmitted as a kind of solution in blood, parental bloods being mixed in the child — a belief carried forth in language: half-blood, pure-blood, blood will tell.

Mendel suggested that hereditary information was a matter not of solution or alloy but of individual units, or characters. Trained in physics, the Austrian monk proposed in 1865 that traits were passed along family lines from parent to offspring in discrete elements, called "factors" (later renamed "genes" by a Danish geneticist). Different factors controlled different traits or aspects of appearance, say, pea color or shape. These factors occurred in pairs, one member of the pair contributed by each parent. The two factors might carry conflicting instructions, in which case the voice of one would dominate. But the other would linger in recessive form, possibly to be rekindled in later generations.

I imagined my own family's marks mixing up and falling out in the neat patterns of Mendel's peas. Facial characteristics recurred in our clan — the Germanic nose with a long swoop and a soft, upturned tip, the multiple moles, the gray-blue eyes dutifully delivered to all four girls. Also, I believed, a tendency to impatience, a love of fresh air and exercise.

But I have since learned that there are brambles along this path. Much heredity does not follow the patterns set down by Mendel. Complex traits, especially aspects of intelligence and behavior, arise not from single genes but from the play of many genes and their environment, both within the body and outside it. Even eye color is more complex than we once imagined, a so-called polygenic trait. The early view that blue eye color is a simple recessive trait has been upended by the brown-eyed offspring of two blue-eyed parents.

Polygenic or not, some characteristics clearly do run in families.

When I'm in a mood, suffering from a bout of hay fever or chilblains that keeps me indoors, I like to page through Victor McKusick's *Mendelian Inheritance in Man,* a two-thousand-page catalogue of every hereditary trait known: dominant, recessive, X-linked. Though the list is now available in a thoroughly modern, up-to-date computerized version, I prefer to thumb the older paper volume for discussions of such traits as:

  the ability to move one's ears, "in males sometimes associated with
    tongue-rolling"
  the ability to smell androstenone
  modification of taste by artichoke, wherein eating an artichoke
    makes water taste sweet
  motion sickness
  male pattern baldness
  the whorl in scalp hair
  dimples

Facial features do tend to crop up in one generation after another, like the large dark mole on the forehead of my brother-in-law, which his wife insisted he have surgically removed, only to find a nearly perfect copy on the forehead of her newborn son. Or those most famous of facial resemblances, the Bourbon nose of the royal houses of Europe and the Hapsburg mouth, that underslung lower lip that showed up century after century in the rulers of Austria and Spain.

In McKusick's catalogue are also rare "mutant" traits that run in families — hairy ears, elbows, nose tip, palms; cherubism, a swelling of the lower face around the third or fourth year of life; distichiasis, or two rows of eyelashes; absence of fingerprints — all reminders that we call contrary to nature what happens contrary to custom. Also listed are mutated genes linked to the susceptibility to rare diseases: acromelalgia (or hereditary restless legs), trembling chin, Tay-Sachs disease, sickle-cell anemia, thalassemia, cystic fibrosis.

Fifty years ago, Linus Pauling showed that the key to innate susceptibility to disease could be traced to an alteration in the molecular structure of a protein, which was, in turn, caused by a defective gene that could be passed right on down the familial line. In this way, a Canadian woman born in 1824 who was affected with aniridia — a defect in a gene that shapes the eye — generated the ailment in whole limbs of her large family tree, with seventy-seven of her descendants suffering from visual handicaps.

Findings of such "disease" genes most often come from studies of families. A gene implicated in dyslexia was unearthed in a study of a large Norwegian family in which eleven of thirty-six members had the disorder. Through such family studies, a gene that contributes to disease susceptibility may be traced to a location on a specific chromosome. The DCP gene, for instance, implicated in myocardial infarction, resides on Chromosome 17. So does the MPO gene, villain in yeast infections, as well as the MAPT gene, linked with dementia, and BRCA1, tied to breast cancer. But to "inherit" the susceptibility to myocardial infarction or dementia is only to hold the gene or set of genes that gives one the potential to develop heart disease or madness — just as to inherit genes for perfect pitch or curly locks is to harbor the potential for musical talent or Milton's "wanton ringlets."

Like most families, mine carries genes linked to common ailments, such as asthma, thyroid trouble, and my own triad of minor ills — ragweed sensitivity, hemiplagic migraine, and Raynaud's disease, or hereditary cold fingers.

But mental retardation?

Beckie's disabilities might have wormed their way into her genome on a sly recessive gene. McKusick lists seven different syndromes of microcephaly linked with recessive genes, including true microcephaly (which in one family cropped up in four of the nine children born of first-cousin parents). Or they might have come from mutations on an X chromosome. Fragile X syndrome, the most common inherited form of mental retardation, arises from a mutation in one end of the so-called FMR1 gene on the X chromosome, a strange stammer of DNA that results in faulty brain development.

Finding a possible hereditary link meant going deep into family history, searching out the branches and twigs of the extended family tree. This was not an easy undertaking in my clan, which possessed no willing elderly remembrancer. My mother's mother had once constructed a webby little pedigree, peppered with holes and festooned with looping curlicues and odd little familial spirals. My father's mother steadfastly refused any such dig into her background and, when begged for a quick pedigree sketch from her memory, produced a crude little shrub topped — only half-jokingly, she said — with Abraham X, horse thief, and his brother Isaac, rapist. The granddaughter would remember what the daughter would forget.

We did unearth a cousin with mild mental retardation, but so radically different was his disability from my sister's, and so distantly related was he to my mother, that there was little possible genetic connection. Beckie's microcephaly might have been produced by her *in utero* exposure to X-rays, the experts said, or, more likely, to a virus. They agreed that my chances of having a child like Beckie were about the same as my neighbor's.

But I wondered about the twigs of disability that might lie hidden among our family limbs. Every genealogical tree has its holes, its secret boughs and branches, the upshot of poor records or unspoken rules about keeping mute on family trouble. Imagine the leaves that might quietly wilt away from official family history — the secret liaisons, discreet separations, the bachelor or spinster howling from some moldy, worm-eaten limb, the wayward running weed that started a secret new offshoot.

The farther back we go, the less we know. Most pedigrees are haunted by tender ghosts of forgotten forebears who are recalled, if at all, as faint mysteries limned in thin anecdotes or brown photos. Still, when one looks at any genealogical spruce or yew, however patchy or lopped, it's pleasing to consider that each of us is fastened to a long string of ancestors — unknown and unknowable in the flesh, perhaps, but faithfully recalled in stuttering voice in our own genomes.

I don't know when I first became aware of science's efforts to weave all the world's organisms into a great family tree, systematically relating one thing to another by way of likeness. I do remember learning from my older sister the neat trick of spotting families of flowers through common characteristics. The crucifers — cabbage, turnip, radish — have a slender seedpod and four petals that form a cross. A violet you can tell by its five petals (the two lateral ones bearded), and by its pistil, shaped like a short beak. Through the mint family, Labiatae, run two flaring lips, square stems, and a distinctive aroma.

When I was twelve or thirteen, my father taught me the Latin binomials for common species of birds: *Dendroica pinus, Dendroica discolor, Sitta carolinensis, Sitta pusilla.* We rose in the early morning to watch birds together, moving quietly in late starlight. Small sounds would hatch from the foliage, not just the normal morning short calls, the hoots, squawks, jargles, whistles, and rasps, but clear strands of real music that seemed to be sung for joy. If luck was with us, we would be jogged awake by the thin fluting notes of a goldfinch or the chromatic cry of a whitethroat. (Later I would learn the surprising fact that stripe-backed wrens so systematically pass their vocalizations down from father to son and from mother to daughter, like a beloved heirloom, that their songs are a highly reliable way of determining family ties.)

Most bird families were a joy to learn: the three little titmice, Paridae; three nuthatches, Sittadae; six swallows, Hirundinidae; seven woodpeckers, Picidae. I found that I could join two birds and see in them one nature; then three birds; then twelve, the faint tracings of individual species converging into major family paths. I remember how the discovery that the natural world had been classified and given two-part scientific names — names that were, in fact, a way of weaving creatures together by natural principles of likeness — was a pleasing revelation.

In the eighteenth century Linnaeus arranged living things in a pattern based on these principles (and freely admitted to being the one chosen to do so, the one whom "God has suffered . . . to peep into his secret cabinet"). I grew to love the Linnaean system, the great animal and botanical divisions, the tidy nests within nests, all the shreds of creation revealed and broken down into pieces, then stitched back together in a great familial tree.

Charles Darwin gave ground for the view that the likenesses Linnaeus observed were quite literally *family* likenesses, that individual species were netted together by threads of ancestry. As he wrote in *Origin of Species:* "All living things have much in common, in their chemical composition, their germinal vesicles, their cellular structure, and their laws of growth and reproduction." In the book's only illustration, Darwin showed variants within a species as branches of a tree. He later explained that the "limbs divided into great branches . . . were themselves once, when the tree was small, budding twigs." All modern species diverged from a set of ancestors, which themselves had evolved from still fewer ancestors, going back to the beginning of life. The relations among all could be represented as a single great dendritic tree.

Darwin himself was heir to assorted intuitions on the kinship of living things. In 1793 William Blake wrote:

> Am not I
> A fly like thee?
> Or art not thou
> A man like me?

A drawing made by Blake the same year shows a caterpillar on a leaf arched over a lower leaf, on which reclines a second simple, cocoonlike form, this one with the face of a baby. The drawing is titled *What Is Man?*

John Clare called flies "the small or dwarfish portion of our own family." That was in 1837, just before the English nature poet was declared insane and committed to an asylum.

"It seems that Nature has taken pleasure in varying the same mechanism in an infinity of different ways," wrote the eighteenth-century French philosopher Denis Diderot. "She abandons one type of product only after having multiplied individuals in all possible modes."

Diderot's contemporary, the naturalist Georges Louis Leclerc, Comte de Buffon, came close to anticipating Darwin. He wrote, "It may be assumed that all animals arise from a single form of life which in the course of time produced the rest by process of perfection and degeneration." A single plan of organization could be traced back from man through fish, said Buffon in his encyclopedic *Histoire naturelle*. (Buffon, who aimed to describe the whole of the natural world in fifty volumes, was so obsessed with his subject, the tale goes, that he evoked from one Madame du Deffand this remark: "He concerns himself with animals; he must be something of one himself to be so devoted to such an occupation.")

Richard Owen, a quick-witted and prescient British anatomist, made detailed observations of the unities underlying life's wild rage of differences. The wing of the bird, fin of the fish, hand of man, he wrote in 1848, were all built according to a single design. These "homologies" were of far greater importance in the scheme of God's grand plan than the minor adaptations that distinguished the organ of one creature from that of the next.

The mind must embrace the whole and deduce a general type from it, wrote Goethe. The German polymath, whose interest in unity grew out of the terrible divisions of his age, possessed the rare ability to leap from science to poetry and back, and — when he was not writing literature — did important work in anatomy, botany, and geology, coined the term "morphology," and even managed to discover a new bone in the human upper jaw. Goethe wrote that all individual organisms have universal tendencies, to transform themselves, to expand and contract, to divide and unite, to arise and vanish, and he claimed to have traced "the manifold specific phenomena in the magnificent garden of the universe back to one simple general principle."

After Darwin, natural historians saw the reconstruction of life's tree as their most important task, scrutinizing the details of shape in adult animals and embryos, the number of rays in a fin, of rows in scales, to deduce the degree of relation. From these, they drew up careful family descriptions, leaving little doubt that fish, amphibians, reptiles, birds, and mammals all descended from a common ancestor. But it would take another century, and a bevy of bright, restless scientists with minds full of physics and with pluck on their side, to unmask the molecular likeness at the heart of life. Strung through all organisms was a single genetic thread made of the same molecular stitches. The difference between a bacterium and a bullfrog lay largely in the order and the sequence of those stitches as well as their total number.

So it was true. The whole code of nature could be written on a thumbnail. Compound it how you will, salamander, German philosopher, algal mat, ginkgo tree, it is but one stuff, the eye of the needle through which all life passes. "Nature uses only the longest threads to weave her patterns," wrote the late physicist Richard Feynman, "so each small piece of her fabric reveals the organization of the entire tapestry."

The findings of the last few decades have made this fact more urgent. Among the genes for tune deafness and hairy palms in McKusick's masterly catalogue are genes for the making of cytochrome C, crystallin, ubiquitin, Hox proteins, and other useful molecules, found not just in members of the human family but in nearly all organisms: worm, ladybird, siskin. When scientists learned to read genes and compare them in separate organisms, their really big shock came on finding the easy way genes from vastly different creatures fell into families, with sequences of stitches so similar that they had to be of common ancestry. These small shared sequences were like family earmarks as distinctive as the square stem of the mint family or the Hapsburg lip.

If a gene has endured almost intact for millions of years in so diverse a panoply of creatures, it likely serves a vital purpose; disable that gene, and the consequences may be catastrophic. Scientists recently scratched around in the yeast genome for sequences like those in certain human genes that, when mutated, are known to cause disease. They found that one in four of these human genes matched a yeast gene, including one involved in cystic fibrosis. And listen to this, William Blake and John Clare: of the 289 known human "disease" genes, 177 have direct counterparts in the fruit fly. (This is good news for research on disease. When scientists find a gene in a worm or a fly that matches a human gene, they can study that gene and its function more easily in the simpler organism than they can in humans, whose genomes are much more complicated.)

Accepted wisdom had it that genes were unique and idiosyncratic, peculiar to particular species. But now it seems that many are close relatives in one degree or another, belonging to one of a few thousand families. Genes shared by the widest range of organisms are probably modern versions of life's oldest genes. Suddenly it's possible for scientists to draw up life's evolutionary tree, to sort out the relations among modern creatures and their distant ancestors, not on the loose basis of appearance — the shapes of leaves and fins, the color of feathers, which are often deceptive — but with tight molecular precision. By drawing the family trees of genes in modern forms of life and running them backward to the genes at the root of the tree, they may even define the nature of the universal ancestor from which all life sprang.

Here is a lesson from the gene-based tree: Life comes in three basic forms — archaea, bacteria, eukaryotes. Organisms made of one cell constitute the majority of life forms, fully occupying two of these three great trunks; we animals are but a slim sprig sprouting from the third. And another lesson: fungi — the broad category embracing mushrooms, molds, and yeast, lumped with the vegetable since Aristotle — are more like animals than plants, sharing more genetic ground with man than with moss or peach tree. (These close relations account for the difficulty of treating fungal infections in humans. So similar are our biochemical pathways that what kills fungal cells tends to harm human cells, too.) The tree of life is looking less like a tidy branching of lily and lichen, fly and dandelion, and more like a thicket, like my grandmother's webby family tree, with more tightly linked ties and binds than we ever imagined.

If unearthing connections with long-forgotten great-grandparents fills one with sweet satisfaction, think of the promise of deeper probing, of uncovering the roots of oneself in deep-lying, distant ancestors of the family — shrew, fish, amphioxus — right down to our oldest, most innovative forebear: the inventor of our brilliant little genetic thread.

2

THE LONGEST THREAD

HERE'S THE HEART of it: a serpentine coil of atoms with an odd stringent beauty, long and skinny, like its name, deoxyribonucleic acid. The molecule of life is made of two chains that twist helically, forming a kind of ladder. Flip the double helix forward in your mind, peer through its center, and you see a small swirling galaxy. DNA has been called a "melody for the eye of the intellect with not a note wasted," a structure of "ingenious parsimony," neat, predictable, orderly.

That is how the models look, the ball-and-stick ladders, the twisted ribbons. I once asked Laura Attardi, a molecular biologist, to show me real DNA. To get it, she simply broke open some cells — they happened to be embryonic stem cells from mice — removed the proteins, and precipitated the DNA with alcohol. She held out a small vial. The DNA was visible to my eye only because it was present in enormous quantities, extracted from millions of cells.

I don't know what I expected, something elegant and beautiful, I suppose, a melody. When I held the vial up to the light, what I saw, with a pang of disappointment, was nothing more than viscid flecks, diaphanous and cobwebby, like mucus.

DNA was first discovered in pus by a young Swiss chemist, Johann Friedrich Miescher, who found the molecule in the ooze of bandages at Tübingen Castle in Germany. The year was 1869, ten years after the publication of Darwin's *Origin of Species*.

When later analyzed, the molecule was found to be made of four chemical bases, among them adenine, found in beef pancreas, and guanine, first detected in bird feces and known, in crystallized form, to give shine to the scales of fish. Before long, DNA had revealed its presence in all plants and animals. But few scientists imagined that so simple a molecule — with its four little bases thought to follow one another in monotonous repetition — could be a candidate for the great secret of inheritance. How could so humble a molecule be the bearer of destiny, the transforming "factors" of Mendel's carefully bred strains of peas, or the hereditary units thought to be carried in linear fashion on chromosomes?

Not only did the DNA molecule seem far too simple for the task of passing traits between generations, but it made up only a tiny portion of the cell, compared with other ingredients: fats, carbohydrates, salts, and proteins. Surely whatever made the myriad species of life and the great gulf of difference between one individual and the next, whatever held the power to create the multifarious pizzazz of life, had to be immensely complex, varied, abundant. The key must be proteins, scientists thought, those elaborate molecules of dizzying diversity that constitute the bulk of living matter. Perhaps DNA was merely the string on which the protein "genes" were strung. But in 1944, when Oswald Avery took purified DNA from one strain of bacteria and fed it to another, even the great microbiologist was taken aback to find that the bacteria with the transplanted DNA were transformed, and that the new trait had been faithfully passed down from one generation to the next.

The architecture of DNA was laid out by James Watson and Francis Crick in 1953 in a nine-hundred-word prose poem packed with meaning, as spare and elegant as its subject. The bases fit together in only one way, clasping consistently, adenine to thymine, guanine to cytosine, a twinned grip as sure as the coupling of two king snakes, so that the sequence of one chain always imposes the sequence of the other. This answered the question of how the molecule can copy itself before cells divide and so pass its message along, uncorrupted, down the generations. DNA replicates by unzipping, the two strands pulling apart, each strand serving as a template for a companion strand. Step by step, the molecule replicates itself until the world becomes richer by another whole thread. In this way we roll along through the generations, like begetting like, squid giving rise to squid, fruit fly to fruit fly, acorn becoming oak rather than, say, maple tree.

The simple structure of the molecule of life also suggested how it conducts the daily business of living through the activity of genes. A gene is a segment of DNA that holds in the sequence of its bases the code for making a specific molecule, usually a protein. Proteins, the building blocks of all living things, are the substances that make familial traits. Each triplet of bases in a gene — say, ACT or AAG — spells out one of the twenty amino acids that form proteins. In the safety of the nucleus, DNA unzips and copies its sequence of bases on a messenger molecule, a species of nucleic acid known as RNA, which journeys out from the nucleus to direct the production of proteins in other parts of the cell.

DNA to RNA to protein. It is a powerful formula, so satisfying and definitive that Crick dubbed it the "central dogma."

I once took a drawing course in the hope that I would learn to illustrate my writings with pencil or ink sketches. Drawing, painting, sculpture, an artist friend had told me, were ways to get at the essence of a thing by understanding how it occupied space. I looked forward to this deep vision and imagined it would come easily. My father is a fine amateur sculptor, and my oldest sister, a talented potter and botanical draftsman. And the start of an artist's work, after all, is not unlike the early work of writing — reaching out for raw material and observing it closely.

In the early classes, which focused on contour drawing — sketching a model without taking our eyes off it — my pencil roamed freely, creating passable work. But when it came to studying master drawings and copying their subtle lines and forms, lights and darks, I failed, as my small errors mounted to produce a miserable drift from the original.

Think of all the blunder and fault in the world. Consider how little exact repetition exists, all the flawed replicas and reproductions floating in the wake of, say, Michelangelo's *Pietà,* copies riddled with small errors and unplanned divergences. The replication of the double helix is a wonder of high fidelity. The DNA in a human cell is made of two sets of three billion base pairs. If each time it copied itself, it made only one mistake in a million (a far better rate than the average professional typist, who misses about one character in 250), that would be three thousand mistakes — a disaster for its owner. When you take into account that DNA copies itself about a million billion times as cells divide during the journey from egg to animal, its ability to make near perfect copies of itself at a rate of about fifty bases a second is nothing short of miraculous.

At the heart of this accuracy is a squad of ingenious enzymes, proteins that orchestrate chemical reactions. One enzyme plucks bases from the cell and "tastes" each one to decide whether it should be added to the dish. If the choice is a poor one, a second enzyme spots the mistake and orders a pause in the building of the chain, signaling the first enzyme to search for a better candidate. Any error that slips through this system encounters a third enzyme, whose role is to detect mismatched bases — in part by recognizing their weaker bonding — and remove them.

It's a brilliant system. Without it, DNA would err every thousand bases or so; with it, the rate of misstep is about one in ten billion.

Though DNA was once considered a supermolecule impervious to harm, the double helix turns out to be — like most things in life — fairly easy to break. It is shocked and disrupted by infections and various poisons, by chemicals and free radicals made during metabolism, by ultraviolet light from the sun, which can fuse together adjacent bases in an exposed skin cell. Each day the DNA of an average human cell suffers five thousand damaging hits from internal heat, which may crack its bonds. Over a lifetime, destructive forces strike every gene billions of times. Sometimes the injury is so severe that it chops the double helix in two. If such a wound goes unhealed, the effect on the cell can be calamitous. Here's the remarkable thing: organisms can sense when their DNA is damaged and delay replication until enzymes can nick out the injured portion and fix it.

Enzymes manage to do their repairs to a molecule that is tortuously twisted, coiled, and packed tightly into chromosomes. If all the DNA in a single human cell were stretched to full length, it would stand three feet high, but so thin is it — seventy-nine billionths of an inch in diameter — and so efficiently bundled that it fits into the space of a cell nucleus just a few millionths of a meter across. To get at wanted parts, the enzymes unbend the molecule, unzip or cleave free one of its bases, and swivel it out to do the necessary repair, one of life's more ingenious tricks performed by creatures from bacteria to blue whales.

It is this stunning ability of the double helix not only to copy itself faithfully but to sing itself back together that allows genes to travel intact from grandmother to mother to daughter and to endure nearly unchanged over eons.

·   ·   ·

It's wonderful to think how things work, to think of the damage that's un-done — and of the damage that's not. Darwin knew nature's anomalies and monsters as saltations; gardeners know them as sports or freaks, the double rose, the branch of crimson blossoms on a white-flowering plant. The word "mutation" was coined by a Dutch botanist, Hugo de Vries, at the turn of the twentieth century. De Vries found a field of evening primroses that had erupted in a riot of varieties — narrow leaves and broad leaves, gi-ant stalks and dwarf stalks — that bred true, a natural version of what Mendel had done with his garden peas. Though the field was plowed under by an unsuspecting farmer, the seed of a new theory bloomed. The means by which new species arise, "new forms that are distinctly different from their parents . . . perfect, constant, well-defined and pure," wrote de Vries, is sudden, unexpected mutation.

Despite the elaborate machinery aimed at perfection, DNA errs. Change occurs along its thread, often through "point mutations" in a single gene — one base substituted for another, one added, one lost — which may subtly affect the protein made by that gene and, sometimes, the trait linked to that protein. A whole block of genes may be lost during replica-tion; a section may stutter or copy itself more than once or lodge on the wrong chromosome. The mistake takes root in a sex cell, is copied from thread to thread, is passed on to succeeding generations, so that a new strain of primrose arises, so that the predisposition to night blindness or gallstones is passed from father to son to grandson, so that chance becomes fate. These small molecular changes underlie many transformations in evo-lution, from a shift in beak shape to the fabulous morphological leap from fin to foot.

This knack for error is a catalyst for newness. Accidents, freaks, anom-alies, are the wellspring of all the devisings of nature. Error opens the door of possibility, turning the course of a bloodline to the east or to the west or straight ahead. Without it, creatures would likely have failed to adapt to all the hard vicissitudes Earth has seen over the ages, the pendulum of climate from desiccating heat to icy cold, the ups and downs of water and oxygen.

Two scientists once took a close look at how environmental pressure affects rates of mutation. Experimenting with the bacterium *Escherichia coli*, they showed that under harsh conditions, strains of bacteria that mu-tate more tend to outnumber strains that mutate less. Under favorable con-ditions, the reverse is true: potent mutators lose the race.

When life is hard, change may help; when it's easy, change may be more threat than benefit. The same holds true for larger species. For finches in the Galápagos Islands, a single year of drought, which makes for tougher nuts, forces an accumulation of errors that drives evolution rapidly in the direction of stronger, wider beaks that can crack a tough nut. Likewise, populations of marine iguanas in the Galápagos shrink in body length by as much as 20 percent in only two years in response to food shortages resulting from periodic El Niños. Small iguanas feed more efficiently than larger ones, so the shrunken lizards survive longer.

Natural selection is the built-in blue pencil, the Yes or No, sampling the change, checking its fit, considering its viability in the whole body and in the environment. If the new thing fails any test, fails to hitch in smartly, or impedes the flow of energy, it withers away. But if in some small way the new configuration raises the level of excellence — allowing organisms to better absorb a nutrient, to disguise themselves, to flee a predator, favoring procreation or survival — it gets a *stet*. Then the change persists as a sweet seed, and a new breed of butterfly or glittering hummingbird shakes itself clean of the dark matrix.

"Error itself may be happy chance," as a philosopher put it. Indeed, said the Russian zoologist Élie Metchnikoff, it would be possible to see man, with his enormously developed brain, face, and hands, merely as an ape's "monster."

This duality of the double helix, this changing and staying, is its real genius. If replication were perfect, there would be little invention, only a planet of onerous repetition; if mutation were unfettered, things would be a chaos of change, nothing with identity, nothing abiding, nothing resembling family. A yin-yang system, one part devoted to stability, the other to reform, has allowed life both to persist and to dream its way into wild variety.

Most of what we know about the gene, its code and conduct, its ability to replicate, mutate, repair its own wounds, we know from *Escherichia coli,* that common, normally benign single-celled resident of the gut — actually from a single strain of *E. coli* gleaned in 1922 from a patient with diphtheria. The microbe's small size, simple nutritional needs, and rapid doubling time (a mere half-hour or so), make it perfect for the study of life's most basic metabolic tasks. In just a few hours it can create colonies of billions.

Microbes may be small, but they make up in numbers what they lack in size. They are the real movers of the world, with the power to sap and upheave nature. Only five thousand or so species of bacteria have been officially identified, but easily that many are contained in a small plot of soil in my yard, a teaspoon of which holds something like five billion individuals. There are more bacteria in the human colon, E. O. Wilson reminds us, than there are human beings who have ever lived.

Still, invisible as they are to the naked eye, it's hard to hold in mind these minute masses. Just as we know that the Earth moves around the sun but still hold to an un-Copernican view of life, so, too, most of us still consider the microbe a penal form, the Alcatraz of nature, as if recognizing the import of the small would diminish our own personal importance. But Pliny said it: "Nature is nowhere to be seen in greater perfection than in the very smallest of her works." The microbial ancestors of *E. coli* were the sole occupants of Earth for two billion years. It was they who gave rise to the three great trunks of life, who invented biochemistry, genetics, sex, memory, communication, decision-making. It is they who, in the last half-century, taught us nearly all we know about the nature of our genes. We may as well look down for kinship.

Not long ago I ran across a fantastic taxonomy in an essay on John Wilkins by Jorge Luis Borges. A true polymath, Wilkins made a modest proposal in the seventeenth century that the world organize all human thought according to the dictates of an analytical language of his own devising. This language parceled the universe into forty categories, among them some so arbitrary (stones divided into ordinary, intermediate, precious, transparent, insoluble; metals into imperfect, artificial, recremental, natural) that they called to mind for Borges another, even stranger taxonomy — recorded, so he claimed, in an old Chinese encyclopedia. *The Celestial Emporium of Benevolent Knowledge* divided animals into:

a. those that belong to the Emperor
b. embalmed ones
c. those that are trained
d. suckling pigs
e. mermaids
f. fabulous ones

g. stray dogs
h. those that are included in this classification
i. those that tremble as if they were mad
j. innumerable ones
k. those drawn with a very fine camel's-hair brush
l. others
m. those that have just broken a flower vase
n. those that resemble flies from a distance

I relish this list, not only for its rhythm and uncanny beauty but for the way it rumples linear ideas of kind, blunts the sharp edge of category, and baffles hierarchy. Here's a reminder that any dividing of life, however useful, is also artificial, reflecting the particular needs of the human mind rather than the realities of nature. It is for the same reasons that I'm drawn to the new findings on DNA, its unexpected behavior and its deep similarities among disparate organisms, from *E. coli* to *H. sapiens*. The molecule of life, far more flexible than we ever imagined, gives rise to newness in myriad ways; its close kinship from one creature to another foils notions of neat sequence, of strange and familiar, same and other, Us and Them.

The conservation of genes across species was revealed not as Darwin's discoveries were — by seeing what others had seen and thinking what no one had thought — but through a kit of new technological tools. Even after Watson and Crick threw light on the structure and function of DNA, its inner world remained cloaked, the details of its features impenetrable, like an unexplored planet. Only with the advent in the 1970s of a new technology — designed, it seemed, to torture DNA on the rack, to slice, stitch, and splice the molecule, to pick out individual genes, isolate them with pinpoint accuracy, make multiple copies, to "sequence" them, read the order of their chemical bases, examine their function, compare them with unlike things — only then did the molecule reveal the hidden teachings of its odd little alphabet. Now, with a test tube, a few chemicals, a source of heat, and a single segment of DNA, drawn from eyelash or drop of murderous blood or hair of a grizzly bear or bone of a Neanderthal, a scientist with a few free hours can make a billion copies of the piece. Then it's possible to analyze that segment, base by base, and compare it with other genetic sequences — either within a single species or between one species and another.

In this way genetic engineering uncovered the existence of families of genes similar in structure, both within an organism and from creature to creature. Such similarities suggested that each gene family had a great-, great-, great-ancestral gene, the original gene of the clan. New members of a gene family arose when nature duplicated that original gene in a kind of biochemical Xeroxing, coopting the copies for something other than their original purpose, modifying them and, perhaps, later rearranging the old pieces in artful assemblage to create something altogether new.

The sculptor Susan Bacik is a master of this art. She works not with raw materials but with found objects. From junk heaps and landfills, attics and yard sales, she plucks old lamp pipes and harp wings, cannibalized microscopes, scales for measuring and weighing, mousetraps, horns musical and animal, shoe lasts, glass bowls, birds, balls, and beads. She touches them up, alters them slightly here and there, releasing them from their cramped and known domain, then welds them together to make beautiful architectural sculptures of startling wit and mischief. The old used-up objects, modified and combined in eccentric hybrids, diverge from their original purpose and are unloosed in the world in wicked new guise.

In her sculpture *Chance in the House of Fate,* an old wire bingo machine, resurrected from a damp church basement and mounted on a high architectural stand, is filled with cat's-eye marbles, the clear kind with the colored swipe inside like a smeared yolk. Beneath the machine is a cup, and crowning the assemblage, a tobacco chopper with a blade like a guillotine. Crank the shaft to twirl the machine, and if chance is with you, a marble may drop neatly into the cup. But the act is made uncomfortable by the fateful presence of the tobacco blade, which, with sufficient jiggling — one imagines with a flinch — might also fall.

Nature works the same sort of alchemy, modifying the old to make the new. Take the globin family of genes, to which belongs the gene that makes hemoglobin, that wondrous bit of molecular beauty uniquely suited to its task: plucking oxygen from air and ferrying it from lung to cell. When animals grew so large that the insides of their bodies were far removed from the outside, they needed a way to get oxygen through the body surface to their tissues. Simple diffusion no longer sufficed. Insects, marine worms, and primitive fish evolved a gene for globin, an oxygen-carrying molecule made of a single chain of amino acids, which could tote the vital substance deep into a creature's interior.

At about the time higher fish evolved, with their need for higher levels of oxygen, the genes for globin duplicated and then mutated, creating a molecule with two slightly different chains and a better ability to take up and release oxygen — the ancestor of our own human form of hemoglobin.

When mammals came about, with their tendency to harbor their progeny within, the globin gene duplicated again, producing yet another chain. This one serves the fetus, creating a hemoglobin molecule specifically suited to transfer a steady flow of oxygen from mother to unborn child. This hemoglobin is stronger, more vigorous than the adult variety, allowing the fetus, like a parasite, to seize oxygen otherwise destined to support its parent host.

On occasion, nature has duplicated and doctored whole genomes. Often the excess DNA was eliminated over generations, but not all of it; in some cases, duplicate genes were diverted to new purposes. In this way the yeast *Saccharomyces cerevisiae* got its ability to brew beer. Our own genome is thought to be the product of several such large-scale doublings early in the evolution of our vertebrate ancestors. We have four copies of many genes or gene families, the Hox genes among them. The human genome, then, is probably an amplified version of a smaller genome, the size of, say, a fruit fly's.

Imagine if we could pull off this trick with flesh, Xerox the original, double the whole shebang, and then select the pieces for keeping. What would I save from such a doppelgänger? Extra hands? Hair? A second slab of gray matter? Twice as many sense cells to smell what is coming?

Newness born through duplication and divergence, through variations on an old theme. It's what we see in a warbler's feathers, in fingers and toes, in the repeating ribs of a scallop, body segments in centipedes, vertebrae in snakes, teeth in a jaw.

Evolution works like a tinkerer, as the molecular biologist François Jacob once remarked, "who, during millions of years, has slowly modified his products, retouching, cutting, lengthening, using all opportunities to transform and create." Darwin delighted in the signs of tinkering in parts of organisms, for they bolstered his theory. "The framework of bones being the same in the hand of a man, wing of a bat, fin of the porpoise and leg of the horse," he wrote in *Origin of Species,* "at once explain themselves in the theory of descent with slow and light successive modifications." Hand, wing, fin, as Richard Owen had earlier noted, all are homologous, related through common descent. So, too, are homologous genes.

When scientists recently decoded the genome of the fruit fly, they were surprised by how few genes it took to make and run the complicated little creature. The fruit fly engages in elaborate behavior, yet it has only twice as many genes as yeast, an organism made of a solitary cell. Apparently, evolution brings complexity to the world, not by creating great numbers of new genes, but by fooling with old ones, like Chaucer reweaving *The Decameron* into some of his *Canterbury Tales,* like Beethoven enriching an old Scottish song.

I like this image of nature moseying around with used objects like a sculptor, pottering with old ideas, working with what it has to breed new worlds.

When scientists learned to read the sequential letters of DNA, another molecular secret emerged. The double helix, long considered pithy and precise — containing just the amount of information required for the form and functioning of a creature, a melody with no wasted notes — turns out to be chock full of junk. Or what was thought of as junk, long nonsensical stretches, sometimes made of the same base or two repeated thousands of times in a droning stutter. These strange genetic wastelands appeared to have no clear purpose, to carry no real information, no instructions for making hormones or hemoglobin or any other protein.

It had been noted that several organisms — sharks, frogs, newts, even beans — possessed too much of the molecule of life, more than was needed to carry on the normal biochemical business of living. The salamander my daughter once caught and held for a day has twenty times the amount of DNA found in her own genome. Field lilies have almost twice that, a hundred billion bases compared with our three billion. The single-celled amoeba, one of the simplest of creatures, has a genome fit for a whale, seventy times the size of ours. But, in fact, only a tiny minority of the helix in amoeba, lily, salamander, man, consists of genes that code for proteins; the rest, 95 percent or so, remained a disturbing blank.

My niece, once told to read a biography of Emily Dickinson and select important passages, chose to highlight the essential facts of the poet's life, not in translucent yellow or blue, but in black, striking the vital and leaving intact only the and's and but's, the stray hat, the mysterious nail in the breast, the unshakable commitment to something — but what? Her mother didn't punish her for defacement but found in the between-pieces of prose and life something worth reading, an oddly poignant poem of its own.

Leonardo da Vinci saw the value of the negative, encouraging his students to paint an "excellent" darkness in order to give meaning to a painting's light, and vice versa, thereby giving birth to chiaroscuro. But most of us are not good at seeing the meaning in dark gaps and intervals. Our senses are tuned to the pulse, not to the pause. When physicists revealed that matter consists mostly of empty space, lonely electrons spinning about immense atomic voids, the world was dumbfounded. So, too, the proposed notion of a wormhole, a region of space that might contain less than nothing — an energy density below zero — remains for most of us an abstraction, one that we can handle glibly but hardly realize. Nor are we better equipped to perceive temporal emptiness, the lag between events and episodes, say, the seventeen-year torpor of our common buzzing cicada, a state of silence that is the insect's normal life.

Scientists were at first perplexed and disturbed by the long, dark noncoding sequences of DNA. It didn't help to note that bacteria and other "prokaryotes," creatures with no nucleus, somehow avoided this problem; their genomes featured wall-to-wall genes. Either these organisms had never been saddled with such molecular excess or somewhere along the way they had sensibly shed it, while we "eukaryotes," with our DNA stuffed into its little nucleus, clung to the baggage.

The silent majority was dismissed as the outcome of excessive gene duplications, the messy detritus of evolution. Indeed, some parts of the junk turn out to be made of those strange genetic stutters, sequences present in multiple copies but never diverted to new use. Others consist of pseudogenes, degenerate remnants of genes that no longer function — old genes for fins, say, or fur — but are carried around in the genome nonetheless, like an old broken compass.

But lately scientists have taken a closer look at the wilderness of junk DNA and found that certain stretches are fecund voids, like Leonardo's darks, full of sequences that may be ungenelike but are nevertheless vital to life, exerting exquisite control over the genes embedded in them. Some such segments stop the ends of chromosomes from fraying like the hem of a skirt, thereby protecting crucial genes from being disrupted; others coordinate and schedule the activities of genes, determining the time and place they are expressed. Some even allow bits and pieces of genes to be more easily shuffled and rearranged into new patterns, which may help evolution in its tinkering.

As if the double helix weren't strange enough, other oddities have popped into view. Genes themselves don't behave as we once thought they did, as neat, indivisible units, each a set of tidy instructions for a single protein, with a fixed job and a fixed location on a chromosome. They have turned out to be decidedly unruly, split, spliced in different ways to make more than one kind of protein, nested, overlapping, redundant, selfish (dedicated to increasing their frequency), even jumpy, footloose, and apt to move around, capable of rearrangement.

The discovery that genes may not be the stable and predictable little beads on a string proposed by Mendel was first made by Barbara McClintock fifty years ago, but, like Mendel's work, was ignored for three decades. Even as a young scientist, McClintock was considered a maverick, a classical geneticist alone in her field, developing idiosyncratic techniques for studying her subject. She had what she described as a "feeling for the organism," an ability to sway outside herself and dwell in other lives, in the homely, intimate details of the maize plants she bred. Her technique was to closely observe the markings and patterns of coloration on the leaves and kernels of her corn plants, study their cells beneath the microscope, and determine the configurations of the chromosomes within. "I found that the more I worked with them the bigger [the chromosomes] got," she said, "and when I was really working with them I wasn't outside, I was down there . . . with them, and everything got big. I even was able to see [their] internal parts."

In the 1940s, McClintock suspected that something peculiar was taking place in her corn; the individual kernels inherited pigments in beautiful patterns that defied Mendel's laws, and the colors changed from one generation to the next much too quickly to be accounted for by slow, gradual "point" mutation. Adjacent parts of a plant were swapping their characteristic patterns, as if one gained what the other lost. McClintock deduced that the beautiful shifting patterns were caused by "controlling elements," which moved from place to place on the chromosomes each time the genome replicated. She then proposed the revolutionary idea that genes are not necessarily fixed in place but can leap from one spot to another on a single chromosome and even from chromosome to chromosome. And, further, that the rate of jumping increases when a plant faces hostile circumstances.

Here was another way that chance might foster newness in the world and life make leaps.

It was the boldest kind of proposal, and it met with puzzlement. Most scientists viewed the stability of genes within the genome as a cornerstone of biology. No one was ready to believe that genes moved, that the DNA of a cell could shuffle its elements. But McClintock was vindicated in the 1970s, when molecular biologists discovered "jumping" genes, or transposons, in bacteria, and, later, in yeast, fruit flies, plants, mice, and men. The DNA of virtually every known species is inhabited by wandering genes able to cut themselves out of one chromosome and splice themselves into another. (Hence their names: *mariner, pogo, gypsy.*) Some of these vagrant bits of DNA may be remnants of viral genes that once wormed their way into the genomes of bacteria, plants, animals. As it turns out, jumping genes dart not only from spot to spot within a genome but from organism to organism, species to species, plant to insect to mammal. A kind of mite transfers pieces of DNA from one species of fruit fly to another with the help of mouthparts akin to the thin glass tubes used by biologists in gene-transfer experiments.

Horizontal gene transfer (so called to distinguish it from the vertical flow that goes from parent to offspring) has likely been a major force of change, influencing the evolution of any number of creatures. Bacteria are especially good at snatching up new genes. A recent study showed that one out of every five genes in *E. coli* came from other microbes. Such pathogens as *Staphylococcus, Streptococcus,* and *Pneumococcus* acquire antibiotic resistance in this way, passing around genes in the form of small ringlets of DNA called plasmids, which enable the bacteria to avoid being destroyed by antibiotics.

It looks as if nature may have been engaged in its own genetic engineering for eons — cloning genes, dicing, slicing, and splicing them. Genes essential to the survival of life have traded hands frequently (which makes life tricky for those trying to construct a universal family tree by comparing genes and tracing their lineage). The human genome is littered with DNA sequences that appear to be the footprints of such ancient "infections." Scientists lately learned that an invasion by a rogue gene in the deep past probably gave rise to our sophisticated immune system, with its ability to invent an almost infinite variety of antibodies. Likewise, it may have been a jumping gene from a viral family that delivered the gift of an essential protein in human placental cells — a fact I learned when I was pregnant. Here's good reason for species modesty: the protein helped prevent my immune system from rejecting my fetus, permitting me to launch my own family.

For all its stability, for all its elegant, conservative, hard-boned structure, the double helix is beautifully limber, full of odd goings-on — which to my mind makes the idea of a single gene, weathering the ride through geological eons to pop up nearly unperturbed in remotely related species, even more astounding.

3

CHANCE IN THE
HOUSE OF FATE

IT IS ONE THING to spell out the thread of life letter by letter, to trace a gene's family and its ancestry; it is quite another to understand that gene's role in life. Most genes are still onion-skinned with mystery; we're only beginning to peel away their meaning.

In the seventeenth century, the English anatomist Thomas Willis wrote that "nature is nowhere accustomed more openly to display her secret mysteries than in cases where she shows traces of her workings apart from the beaten path." I think of my sister Beckie's deficits illuminating the nature of sensibilities outside the intellect, unexpected moving powers of spirit and emotion. I think of the losses neural and mental — aberrations of speech, memory, perception — that have shed light on the normal cognitive powers of the human brain.

Williams syndrome, for instance. Every so often, in one out of twenty thousand births, a child comes into the world with this odd constellation of features: elfin face, with upturned nose and wide mouth, full lips, and small chin; heart defects; and, by the standard of IQ tests, mental retardation — but also with special qualities of mind preserved and enhanced, including extraordinary powers of sociability and language. A child with Williams syndrome may not be able to tie her shoes, use a fork and knife, cross the street, read, write, or draw, yet be capable of engaging adults in long, involved conversations, telling imaginative stories, composing song lyrics, using grammatically complex language and rich vocabulary. Ask her for a list of animals, and out comes ibex, musk ox, manta ray, newt. The rare disorder — which results from a missing piece of Chromosome 7, containing sixteen to thirty genes, including the gene that makes the protein elastin — has pointed a path to understanding the still inscrutable genetic origins of language and sociability.

The key to normal being is often found in the outskirts and extremities of nature, the deficits, excesses, and aberrations. This is true for genes, too. Treasure your exceptions, urged an English geneticist. Most of what we know of genes and their purpose comes from studying how nature responds to having its tail twisted, often in model organisms — fruit flies, roundworms, yeast, zebra fish, mice — which can be grown, mutated, and analyzed with speed and precision. Such mutants can reveal what a gene does by showing what happens when it's knocked out or disabled through radiation or chemicals, a method geneticists call "wreck and check."

So many new genes have been unmasked in this way that scientists have been engaged in Adam's task, reining in the molecular chaos with a net of witty, irreverent names. Drawn from poetry, mythology, fairy tales, movies, cookbooks, varieties of wine and cheese, the names often describe what an organism can or cannot do after a gene has been disabled: *cheap date* (a mutant gene that makes a fly easily intoxicated), *tin man* (one that robs a fly of its heart), *chico* (Spanish for small boy, a gene that reduces the size and number of cells and leaves a fly less than half the normal size), *cyclops, fruitless, quagmire, quick-to-court, killer of prune, nuclear fallout* — all names that treasure the exception.

While it may be the gene that suffers the slings of change, the illuminating accident occurs in the product of that gene, most often a protein.

All the flesh in the world is made of proteins, all the plasma and pods and platelets, the spines of hedgehog and prickly pear, the skin of rhino, opossum, and pachyderm, all the chitin, chlorophyll, collagen, hemoglobin, hair, horns, bones, eyes, and brains. This is what makes it worthwhile for one member of the family of life to eat another. The fiber and flesh of plants and animals give us the raw material we need to make our own highly specialized suite of proteins.

Genes exert their effects largely by specifying which proteins are made in a cell. A typical mammal cell contains about ten billion protein molecules of perhaps ten thousand different kinds. They are its true wizards, charged with the tasks of metabolism, digestion, excretion — all the multitudinous coordinated activities that direct and sustain life. Some proteins lend strength to the cells of skin, muscles, bones, and tendons to give the body form. The structural protein elastin, missing in people with Williams syndrome, builds elastic fibers in arteries, lungs, and skin. A pair of proteins, myosin and actin, expand and contract the 228 muscles in the head of a caterpillar, and a team of proteins control the thousands of operations by which the sharp jay strikes its caterpillar prey. Some proteins act as messengers or the receivers of messages from distant cells. Others speed up biochemical reactions that otherwise might take years, let sperm into egg or forbid its passage, help a baby's brain cell find its way from eye to visual cortex, or prepare DNA for a necessary genetic event, such as the making of a new protein. Some proteins grab hold of the double helix and squeeze, twist, or tweak it; these are the means by which one gene "regulates" another, switching it on or switching it off, thereby influencing everything from the efficiency of the respiratory system to the sex of an infant.

Just as our millions of words are made of the twenty-six letters of the English alphabet, and the thousands of characters in Chinese writing are composed of seven basic brush strokes, so all proteins are made from various combinations of twenty amino acids — eleven of them made in the body, and the others supplied by amino acids in food. Proteins are constructed in splendid little structures called ribosomes, which are themselves made of fifty-four different proteins. The ribosomes respond to instructions from DNA and — with the help of the go-between molecule RNA — make amino acids, which they link up, one after another, converting DNA's sequence of bases into a chain of amino acids. Although proteins start in linear form, unreeling from the ribosomes as long thin strings of these amino acids, within seconds of birth they crumple into a bewildering origami of spirals, pleats, zigzag sheets, and confetti-like helices — bewildering, but highly precise. Even a relatively petite string of a hundred amino acids could in theory fold into as many as 10^{100} different shapes. That's a google, a very large number indeed, bigger than the number of atoms in the universe. Yet a given protein nearly always snaps reliably into its designated shape.

On a windowsill in my study I keep a display of seashells gathered from my travels, an elegant high-spired Terebra shell, a radial-ribbed Cardiida, the blunt, utilitarian rectangle of the razor clam. I started the collection after I stumbled on a book about the natural history of shells, an elegant exploration of shell geometry and geography, by Geerat Vermeij, a biologist blind since he was four. I admire the shapes of shells, as Vermeij does, the spirals, ribs, knobs, spines, the polished interiors smooth to the touch. Tactile clues are Vermeij's only access to the shell's fine form and sculpture; it is these very bumps and ridges, he says, that reveal and record not just the everyday events of the shells and the strange circumstances that sometimes mark their lives — the losses and the triumphs — but also the evolutionary interplay with other species and the crucial biological consequences of form.

In the tiny, blind world of proteins, too, shape is everything. Proteins work by touch, by bumping, grasping, and clasping other molecules. It is their particular architectural details that give them their functional power; their bulges, bumps, hollows, and pleats determine whether a cell will metabolize nutrients, capture passing hormones, make new DNA or new proteins, recognize bacteria, receive signals from the bloodstream, grow normally or become malignant, make pleasure or pain, wonder, love, hunger, adoration, hatred, terror.

I once saw a floor-to-ceiling model of hemoglobin, the protein that picks up oxygen and unloads it in our bloodstream. Even at that scale, the molecule was an unreadable tumble of sticks and balls. The great chemist Max Perutz labored for twenty-two years to map out its ten thousand atoms. When all the pieces finally came together, Perutz stood, shocked, before the twisted mass: "Could the search for ultimate truth really have revealed so hideous and visceral-looking an object?"

Hemoglobin rates as relatively simple in the pantheon of proteins, some of which consist of hundreds of thousands of atoms. Compared with cool, rhythmic DNA, proteins are messy, globular, arrhythmic, with long untidy streamers of atoms trailing from their edges.

Though scientists know that the order of amino acids dictates how a protein will fold up into an active, three-dimensional body, they still can't predict what structure will emerge. Other factors play in. There are helper molecules that push or pull a new chain into shape. When scientists first discovered these so-called chaperones, they thought their main job was to keep young amino acid strings from randomly shacking up with one another. A cell is a crowded place, with billions of protein molecules floating about and new ones unreeling at a rate of two thousand a second, so there's plenty of opportunity for wayward coupling. (Making proteins at such high speed results in large amounts of error and waste. Close to a third of all new proteins are destroyed within minutes of their birth because they haven't folded into their proper shapes.) Chaperone molecules, it turns out, are sculptors, urging proteins to assume their assigned forms, nudging a helix here or shifting a loop there to help create the pocket inside a hemoglobin molecule that embraces oxygen or the protruding arm of an immune cell that reaches out to grab a pathogen.

Only when a protein is wadded up in this way can it jump to its designated task. Misfolded proteins may huddle in useless clumps, damaging the cells around them. These misfits are considered the culprits in wrinkled skin, hardened arteries, and such diseases as cystic fibrosis and Huntington's disease, and they may be responsible for the plaques and tangles found in the brains of people like my grandmother, who suffered from Alzheimer's disease. This is one reason scientists are so keen to decipher the details of protein shapes.

Protein scientists aim to parse the structure of their chosen molecules atom by atom, just as Max Perutz did with his hemoglobin, determining the exact position of thousands of individual atoms. In an effort to visualize their quarry, they gesture and scribble, drawing ball-and-stick diagrams on blackboards or constructing them from plastic sets (black balls for carbon, blue for nitrogen, white for hydrogen). They talk about "walking" down the backbone of a protein from one amino acid to the next. But so extremely small are most proteins that those who study them work completely in the dark.

We are too big for our world, says Asher Treat, a biologist devoted to the tiny mites that infest moths and butterflies. Treat prefers the pocket wonderland beneath a lens, "where a meter amounts to a mile" and life slows considerably. This was the sentiment, too, of Anton van Leeuwenhoek, the naturalist from Delft who had that first breathtaking look at minute life through the bright well-hole of a powerful microscope and thereby stole heredity from the exclusive province of philosophers. Leeuwenhoek delighted in the secret parts of nature, the swarming forms so small that the naked eye could not discern them, even those extracted from the white matter "that sticketh or groweth" between the teeth.

I then most always saw, with great wonder, that in the said matter there were many very little living animalcules, very prettily a-moving. The biggest sort had . . . a very strong and swift motion, and shot through the water (or spittle) as a pike does through the water.

The notion that a drop of water could hold thousands of gyrating, dancing creatures, with sense, energy, life, scandalized many of Leeuwenhoek's contemporaries, Buffon among them. Though a man of wide interests who came close to anticipating the work of Darwin, Buffon did not wish a world of invisible beings, "wondrous" or not, and considered the idea of animalcule savages wiggling and biting in a water drop an insult to mankind. He dismissed Leeuwenhoek's microscopic observations as the product of an overzealous imagination.

This did not stop Leeuwenhoek from building more than five hundred microscopes to explore the least dimension of living things: the animalcules like tiny dancing gnats or flies, the sex cells, the crystalline lens of the eye, the striation of the muscles, the red corpuscles of the blood. All things visible to the naked eye paled in comparison with these secret minutiae. The invisible had become visible, and the revelation was startling.

It's true. Under a powerful microscope the commonplace is revealed in hallucinatory detail. A pinch of sand becomes a tumble of bold, brilliant crystals. A fly's leg is transformed from a snip of black thread into a thick, hairy limb, intricate with claws and an ingenious gray suction pad for walking upside down. There is in this magnified world, wrote Vladimir Nabokov, "a kind of delicate meeting place between imagination and knowledge, a point arrived at by diminishing large things and enlarging small ones, that is intrinsically artistic." Nature itself defamiliarizes the world in this way, freely makes big things of small things and small things of big.

In Leeuwenhoek's day, the most powerful lenses were capable of magnifying objects several hundred times to show the silent luminous beauty of a single cell and the organisms that swim between our teeth. These days a good optical microscope can magnify more than a thousand times, illuminating the larger chromosomes inside the nucleus of a cell. While light cannot probe things smaller than its own wavelength, a beam of electrons substituted for visible light in an electron microscope can magnify 500,000 times, revealing the fine details of a cell's organelles and even the presence of large molecules.

Protein scientists descend in scale even farther than this, down past nuclei and chromosomes and macromolecules, farther down to that world where chemistry shades into physics, to the atoms that make up minute molecules on a scale measuring a billionth of an inch. Only there can they scrutinize a protein's intricate nubs and clefts. Their main clues to shape are the laws of chemistry that define the strength and geometry of bonds between atoms; among their chief tools is X-ray crystallography.

Though proteins seem chaotic, they are so precisely shaped that millions of molecules will align perfectly, one atop another, to make a protein crystal — an easier target of study. X-rays aimed at the crystal scatter in unison when they hit the electrons in the stacked atoms, creating "fingerprints," or patterns that reflect the molecule's atomic structure. Imagine being blindfolded and asked to describe, without touching it, an automobile sitting in front of you. You are given a bushel of fifty thousand pebbles to throw at your subject, told to calculate the angle at which the pebbles bounce off it, and, from this, you are expected to determine the car's precise shape.

The biochemist Ann Ferentz uses an even less direct way of seeing. The protein she studies — a molecule made by cells in response to ultraviolet light — doesn't crystallize, so she dissolves it in water and salt, then sticks the solution inside a nuclear magnetic resonance spectroscope, something like a big magnet. This provides her with bits of disembodied information about the location of hydrogen atom pairs in the protein. She measures the distance between the atoms in a pair, and, through a series of calculations and inferences, circles in on the protein's shape. The data are hard to interpret, and the picture is incomplete, so Ann often ends up at that delicate meeting place of knowledge and imagination. To grasp the whole shape, she thinks herself inside her molecule, as Barbara McClintock did with her maize, imagining what it's like to be its atoms, aiming for an almost muscular feeling for the many electrochemical forces — some local, some long-distance — that bind the atoms together or push them apart, bunching up a protein chain so that remote parts end up cheek to cheek. She has that ability possessed by certain geologists who, by some clairvoyant extension of senses, can see in their heads, as if by aerial photography, the meandering roots of a whole watershed — but in Ann's world, the scale is reversed.

If you analyze the shape of hemoglobin, what you get is 574 amino acids in two pairs of chains (two alpha, two beta), each arranged in an exact relationship around a disk-shaped, dark red heme molecule, which gives blood its color. At the center of the disk is a single atom of iron, which grips oxygen. When a hemoglobin molecule arrives in the lungs, its iron atom binds to oxygen and carries the precious gas to tissues. Now, if you remove the sixth amino acid in one of the chains and put in its place another amino acid, what you get is weird folding, an odd pocket, distorted red blood cells, and sickle-cell anemia. Tinker again. Remove the atom of iron from deep within the molecular pocket and in its place put an atom of magnesium. What you have now is chlorophyll.

But here's a paradox. While proteins may seem singularly touchy, requiring little to undo them — a chance mutation, perhaps, or, as cooks know, a touch of heat — they are also robust. A protein "denatured" will, in a matter of seconds, snap back to its native shape. (Up to a point. The denatured proteins in a fried egg will never regain their original shape.) And though a certain swap of amino acids can undo hemoglobin, the protein tolerates other such changes. Victor McKusick's *Mendelian Inheritance* lists hundreds of variants of hemoglobin with single amino acid substitutions, each graced with a name — Fitzroy, Flatbush, Zagreb, Abraham Lincoln, Beijing, Beirut, Belfast, Coventry, Cowtown — and most of them capable of snapping up iron, just as they should.

Because protein molecules are so precise, so intricate, irregular, and wildly diverse, scientists supposed that they must be made of hundreds of thousands, even millions, of idiosyncratic parts. But in analyzing the shapes, they've found a surprise. Widely different proteins are made of identical parts. These parts are structural "motifs," discrete little blocks of amino acids that reliably fold into patterns of helices and sheets. The motifs carry such names as kringle (for the Danish pastry it resembles), apple, kunitz, link, zipper, zinc finger, forkhead, sushi domain, and homeodomain. The helix-turn-helix motif, a protein shape that binds neatly with DNA, has been found in hundreds of proteins that switch genes on and off in organisms from bacteria to humans. The key to whether a single tweak in a protein's string of amino acids is inconsequential or radical, harmless or lethal or useful, often depends on whether it occurs in one of these critical regions.

These helices, kringles, and links are like miniature proteins within a protein. And they may be just that: ghostly vestiges of the earliest proteins formed when life had barely begun.

From the twenty amino acids that emerged during Earth's first thousand million years, the earliest kinds of life fashioned a set of basic protein shapes that did their jobs with subtlety and precision: arms that would catch, jaws that would open and close, helices that would squirrel neatly into the grooves of DNA. "These early shapes were so vital, so good at what they did, that they made themselves indispensable," the evolutionary chemist Russell Doolittle told me. "They became keystones that would persist down the ages and now appear in proteins in all cells, including our own, much as they did in our primitive ancestors a billion and a half years ago."

It is these kringles and fingers and other old, archetypal protein parts that may explain the existence of some of those "junky" stretches of gibberish in our genomes. The idea is this: our genes contain units called exons, which code for protein motifs. One exon may code for a protein's binding site, and two others for its armlike side chains. The nonsensical junk between the exons, called introns, would allow these pieces to move as discrete units from one protein to another, to be shuffled and arranged in new patterns without losing their function.

At his laboratory at the University of California in San Diego, Russell Doolittle compares proteins from different organisms to see which may be related, and how far back the relations go. With the help of a computer program, he aligns the amino acid sequences, then looks for those that are similar enough to suggest a common ancestry.

Take cytochrome C, a ubiquitous protein with the task of shuttling electrons during cell respiration and photosynthesis. A sequence of 104 amino acids describes our own version of the protein, the same 104 possessed by the chimpanzee. A rabbit's cytochrome C shares 92 amino acids with ours; a duck's, 87; a rattlesnake's, 84; a moth's, 68; and yeast's, 38.

"As a rule of thumb," Doolittle told me, "any two protein sequences that are more than 25 percent identical are likely to be descended from a common ancestor." But even proteins whose sequences are less similar may be related, though very distantly. "In these 'twilight zone' cases, when you're comparing two proteins that shared a common ancestor in some very remote time," he explained, "to detect relations you have to compare the three-dimensional shapes of the proteins."

Leghemoglobin, a protein in plants, shares only 15 percent of its amino acids with our hemoglobin — close to what one would expect from chance. But the shapes that make up leghemoglobin are nearly identical to those of hemoglobin. Though the sequences of amino acids in the proteins have changed, some solid, ancient core of shape remains.

Here's the surprise. Even when all "memory" of an amino acid sequence is lost over millions, even billions, of years of evolution, the overall three-dimensional structure of a protein can remain unchanged. "Imagine you're building a house of Legos," explained Doolittle. "It has a certain distinctive architecture, an L-shape, with a flat roof, say. Now replace the Lego pieces one at a time. You can change every piece of Lego, and the house will still look the same. As long as the substitutions are small and conservative along the way, the original structure is maintained."

When I was young, I wondered why human hearts pump thin blood rather than something syrupy and slow-running, like sap; why all other creatures were clothed by nature with shells, bark, wool, hide, feathers, scales, while we are shaped with the soft side out, vulnerable to thorns and rain.

The answer, of course, is that we are sheathed in the shapes of the past, the skulls and shells and skeletons, the kringles and fingers, which make fate, preventing the possibility of moving in certain directions and closing behind us the gate of conceivable gifts. "Nature is what you may do," wrote Emerson. "There is much you may not." Our blood cells took the salt habit from the sea they evolved in; our bone cells, the habit of lime. Our arms swing in opposition to the swing of our legs, because four billion years ago our piscine ancestors, with their two pairs of fins, wiggled over the warm mud by throwing their bodies in an S-curve.

We are born with the ability to make close to two hundred different sounds — hisses, hums, squeaks, whistles, and pops — but in spoken language, most of us use only a few dozen. Still, with this narrow range, we create nearly limitless sentences with radically different meanings. So, too, some birds have only a few song motifs, which they rearrange to produce an impressive repertoire. With proteins, nature does the same thing, using a limited selection of ancient protein shapes, known quantities that work, to create a staggering diversity of molecules. The solid three-dimensional structure of these basic protein shapes is the robust platform on which nature makes modifications. Put a skylight in the flat roof of your Lego house, sure, and a deck off the back — the house will hold. Life is bent on diverging, but it builds on what is already there; in the case of proteins, a set of shapes older than anything brought to air by pick or shovel, and rock-solid stable.

Yet here's another paradox.

When Ann Ferentz talks about proteins, her hands dance and flutter birdlike, fingers closing, opening, bending, to summon form and vivify the flat motionless symbols on paper that convey only a shadow of her meaning. Proteins do not escape the incessant shiver and wobble of all life; the bonds that hold together their atoms are less like glue and more like springs, able to stretch and squeeze, expand and contract. So flexible is hemoglobin as it pulses between its two shapes to do its work that Max Perutz called it a "molecular lung." In motion, too, are the individual atoms within a protein. They quiver, shake, and vibrate in a time span measured at trillionths of a second.

The genius of proteins, like that of the genes that make them, lies in this duality, this remarkable ability to hold shape and yet change it.

When I'm feeling stuck and lumpish, dropped like a stone in a personal rut, I like to remind myself that fixity in nature doesn't exist. The chair in which I'm slumped spins with the Earth at eight hundred miles an hour, whirling with the whirl of the solar system, which is swinging around the Milky Way galaxy, which speeds outward with the expansion of the universe itself.

Going the other way, down, the tendency toward motion appears on all hands: the river's leaping force locked in my veins, my eyelids fluttering even as I sleep, the ribosomes chattering away by the billions in my cells, spinning out thousands of proteins a second, proteins that are themselves composed of agitated atoms flicking, quivering, and kinking to brighten my blood and stay my form — long enough, at least, for me to pass along to the next generation the solid little kernel of my being.

PART II

GENERATION

A body of men, animals, or plants having a common parent or parents; the act of producing offspring; the process of coming into being. From the Indo-European *gen-*, to give birth, beget.

4

SEEDS OF INHERITANCE

NOT LONG AGO the newest marvel in biology, covering all the front pages, was the cloning of a sheep from an udder cell of a six-year-old ewe. The new, same sheep, christened Dolly after Dolly Parton, leaped into nearly all conversation. The phenomenon of cloning wasn't new in nature, of course. Amoebas are masters of the technique, and dandelions, too, as well as that vast, sprawling *Armillaria* fungus that infected so many miles of Michigan forest. The hoe-split earthworm grows into two identical worms. Whole willow trees grow from cuttings. In just this way, farmers and gardeners propagate garlic, grapes, bananas, apples, and sugar cane. In fact, the word "clone" is derived from the Greek word for "twig."

But the cloning of a mammal from an adult cell *was* revolutionary. Each one of a sheep's udder cells, like most of its specialized cells, carries a complete set of genes, all the information needed to create a whole sheep. But once a body cell becomes specialized to perform its particular duty, its fate has been sealed. It has presumably given up its ability to start over and create a whole new being; that remains the sole province of embryonic cells.

Here was the remarkable thing. Scientists working in a lab in the hills of Scotland had tricked an adult cell, one that had taken on the job of mammary cell, into believing it was an unspecialized embryonic cell, and then coaxed it into starting life all over again. The feat took 434 tries, but it finally worked. By a jolt of electric current the adult udder cell was fused with an egg cell whose nucleus had been removed. The "denucleated" cell was "renucleated," and the genes of the udder cell were seduced into acting as if they were in the same state as when sperm first fertilized egg. Once the newly created egg started dividing, it was implanted in a surrogate mother sheep, and several months later Dolly was born.

Goat, cow, pig, and mouse soon followed, then the sheep Cupid and

Diana and the monkeys Neti and Ditto. Now it seems likely that a version of the technique may soon be tried with humans.

The cloning of mammals has popped to the top of the list of Things Science Can Do, spurring heated arguments over ethics and the boundaries of research, whether scientists should have the power to create and control life. The queasiness created by the news is dispelled only a little by the knowledge that most scientists support cloning primarily as a method for making medicinally useful human proteins or cells or tissues for use in transplantation — supplying people with new tissue of exactly their own genetic type and thereby avoiding the risk of immune rejection.

Nor are fears necessarily soothed by the understanding that even if someone does figure out how to clone a human, identical genes do not make identical people. Nothing is so unlike as two peas in a pod. Careful cloning of single plant cells produces diverse offspring. Genetically identical bacteria have unique characters, shaped by chance fluctuations in a few of their key molecules. As for humans, identical twins unfold in the womb in different ways. And once those twins emerge into the slapdashery of life — diseases, tropisms, obstacles, parents, family, schoolmates, friends, the Bronx or Oak Park, Switzerland or Serbia, not to mention variations in earth, water, and air — their bodies hopelessly diverge. And their minds. The network of neuronal connections that make up the individual brain depends on internal thoughts and on sensory information from the outside world — a quick conversation, a note of music, an image of war — which act to strengthen or weaken particular neuronal paths. The mind creates itself through its own unprecedented and unrepeatable history.

Still, this cloning business is fertile ground for speculation and for nightmares of unnatural choices unnaturally made, of parents selecting the genetic identity of children for reasons of vanity or for practical ends, of whole families of doppelgängers created from the DNA in a flake of skin or snippet of hair, like so many olive shoots around the table, of armies of "ideal" soldiers, of uniqueness, sex, and death eliminated — at least theoretically — and, with them, human dignity.

All the fuss about cloning has blinkered some items of new knowledge about an older surprise that I find no less monstrous or miraculous: the random, chancy coupling of sperm and egg and their strange play in heredity.

Sex cells are not what we once thought: the egg a passive, slumbering beauty; the sperm, a cheap messenger. Nor is their joining a simple story of cooperative consummation, of happy meeting and mixing of genes, but, rather, a Byzantine tale of intrigue, conspiracy, fierce competition, conflicting interests, delicate negotiations, and — only rarely — success.

Consider the egg, life's most ancient specialized cell. "I think if required on pain of death to name the most perfect thing in the universe," wrote the naturalist T. W. Higginson, "I should risk my fate on a bird's egg." In Greek history, a primordial egg laid by a polymorphous monster broke open to give birth to both earth and sky and, later, to gods and mortals. The distinguished physician William Harvey placed on the frontispiece of his great *Exercitationes de Generatione Animalium* of 1651 an illustration of the hand of Zeus clasping an egg, out of which arise reptile, insect, bird, child, and this motto: *Ex ovo omnia.* "See this egg," said Denis Diderot. "It is with this that all the schools of theology and all the temples of the earth are to be overturned."

Self-contained as the moon, as an O, the egg is one of those deeply conservative cells by which we vertebrates are linked to slimy, spineless things; it has conveyed us all from one generation to the next for hundreds of millions of years with little fuss.

I once stumbled on a wonder in the halls of a museum, a replica of the egg of *Aepyornis,* a giant flightless elephant bird that lived in Madagascar during Pleistocene times. The eggs of *Aepyornis* were encased in shells so heavy and stout, they were used by ancient aborigines as water bowls. Each tipped the scales at a thousand pounds. To think: this giant was neither more nor less than a single cell, one of the largest that ever existed.

Brought up as I was on the familiar modest oval made by a hen, it's hard to consider *Aepyornis*'s ovum within the rubric of egg. So, too, the eggs of the blind snake, with its unusual rodlike shape, or the slimy, jelly-like masses of millions of pearly eggs that the horseshoe crab pours out in cold fecundity, or the long sticky strands of eggs released by a female octopus in her one fervent shot at reproduction. Or, for that matter, my own speck of an egg, barely visible to the naked eye, one tenth of a millimeter in diameter and yet responsible for carrying on the family legacy. But the enormous gulf in size between my egg and that of the elephant bird is only a matter of the way the embryo is nourished. Bird eggs are swollen with yolky proteins. My egg has no shell and not much of a yolk, my line having evolved the ability to hold and protect eggs inside while the embryo develops in the safe, warm, nutritious environment of my reproductive tract.

In nearly all species, the egg looms large in relation to other body cells. This is because of the heavy load it carries — proteins, fats, minerals, nutrients, and powerful chemical signals that lure sperm, literally egging them on. The egg also holds ultraviolet filters to shield the embryo from the sun's harmful rays and a stash of foul-tasting chemicals to protect against predators. It has special enzymes that repair the damaged or defective DNA in sperm, as well as molecular signals to start the embryo growing and others to direct the fate of its cells. All of these riches are stored in forms that can be used in precisely the right way at precisely the right time to direct fertilization and development.

Not long ago, the biologist Hubert Schwabl discovered that female songbirds bestow upon their eggs a surprise gift: a powerful dose of the hormone testosterone. The dose varies according to the order in which the eggs in a clutch are laid. Somehow, the songbird increases the amount of testosterone with the order of laying, stepping up the vigor and competitive abilities of younger nestlings, perhaps giving them an edge on their older siblings.

Allan Spradling, an embryologist at the Carnegie Institution of Washington, D.C., takes the eggs of fruit flies apart, gene by gene, to see how an egg learns to be an egg. The eggs of fruit flies, it turns out, are produced by the communal effort of more than a dozen cells. One lucky cell becomes the oocyte, which gives rise to new life; the remaining fifteen become nurse cells, serving the oocyte by giving it the raw materials it needs to function. Spradling has seen hints that these nurse cells may be providing eggs with high-quality cellular organelles, particularly mitochondria, the cell's little powerhouses. "'Good' mitochondria somehow are recognized and moved into the future egg," he explains, "while damaged mitochondria are likely removed by a reverse journey." The communal approach to raising an egg appears almost identical throughout much of the animal kingdom — another shared inheritance of great importance.

With all these riches produced so laboriously, it's no wonder that eggs are costly and rare and that they age over time.

Now consider the sperm. In the Middle Ages, the immensely popular *De Secretis Mulierum,* or *Women's Secrets,* defined sperm as "nothing other than excess food which has not been transformed into the substance of the body." Earlier, the Yemeni midrash suggested quite otherwise. A sperm collects within it all the parts of the body:

> It originates in the brain and then it settles into the arteries. The arteries introduce it into the spine and the spine divides it up among the organs; from the organs [it goes to the sinews], and the sinews bring it to the places where the water is stored. It does not leave the body until it mixes with all the inner parts of the body. For that reason there is a sensation when it leaves, and it is sour.

When Leeuwenhoek first saw sperm through his microscope, he considered them parasites living in the semen ("spermatozoa" means sperm animals) and assumed that they had nothing to do with reproduction. This was also the belief of Lazzaro Spallanzani. Despite his experiments in the late 1700s showing that toad semen stripped of its sperm would not fertilize eggs, Spallanzani held firmly to the belief that the key to life lay in semen's viscous fluid, not in its spermatic animals. It was a Swiss zoologist, Hermann Fol, peering though a microscope at a starfish egg in 1876, who was the first to see that fertilization was a matter of sperm penetrating egg. The biologist Oscar Hertwig, witnessing the same event in a sea urchin that very year, noted that only one sperm enters each egg, and that the nuclei of the two cells unite.

Sperm are not, as once believed, a mere splash of nucleus wrapped in a protein coat — uniform, simple, devoid of resources. Each sets out with some splendid equipment. There's the head, which holds a nuclear package capped by a powerful packet of enzymes used to lyse the outer coating of the egg, a neck stuffed with mitochondria to provide the sperm with energy, and a long tail of brilliant design to help it travel long distances under difficult conditions — all necessary gear. A sperm's tortuous trek from epididymis to egg is unpredictable and dangerous. During ejaculation, it will speed through the male tract at more than eight thousand body lengths a second, the equivalent of a person moving at thirty-four thousand miles an hour. Once out, it may suffer shipwreck during intercourse and leak from the vagina or be ambushed by white blood cells and summarily destroyed.

Good propulsion is of the essence. The sperm's tail shaft, or axoneme, is made of two central microtubules surrounded by nine outer ones, a construction both ancient and efficient, beautifully suited for movement. This nine-plus-two pattern turns out to be universal in such flagellar shafts wherever they are found in nature — in the sperm cells of buffalo semen and those of banyan trees, in the tails of swimming protists, in the cilia that clear our windpipes of debris and those that gently urge the egg from ovary to womb. In mammals, the sperm has another bright bit of engineering: dense fibers between the axoneme and mitochondria protect it through the stress of its journey and prevent its head from being whipped around violently.

Nicely dressed as he is, courageous and questing, still the sperm suffers in comparison with the egg. The sperm of *Aepyornis* was invisible to the naked eye. A human egg is 250,000 times the size of a sperm, a marriage of marten and mountain.

It wasn't always this way. In the red morning of life, there was no great size gulf between the sex cells, no sperm or eggs at all, for that matter, no males or females. There was reproduction, of course. Simple organisms floating about in some long-forgotten sea reproduced by cloning or fission (self-division) as bacteria do today — a good way to create a new generation and pass on one's genes intact. There was sex, too, in the bacterial sense, the swapping of genetic information by "conjugation," the opening of a hole in two cells through which they exchange genes but don't fuse or swap organelles and don't create a new individual.

Then, about a billion years ago, reproduction and sex somehow stole a ride upon each other. Perhaps one cell engulfed another, the genes of the two creatures merged, and the cell divided to form a new being different from either of its parent cells. Perhaps the commingling of genes gave the changeling an edge by purging harmful mutations and collecting good ones that allowed it to resist parasites and other infections.

The first sex cells may have been interchangeable and of roughly the same size. By chance, some may have been slightly bigger than others and stuffed with nutrients, an advantage in getting progeny off to a good start. Perhaps some were smaller, faster, good at finding mates. As organisms continued to meld and join their genetic material, the pairing of a larger cell with a smaller one proved an efficient system. Over time, the little rift between the sexes widened, as did the strategies of male and female for propagating their own genes. The male's overriding interest was to fertilize an egg before a rival male did; the female's goal was to seek the best genes for her offspring and make sure her egg was fertilized with one sperm only — fertilization by multiple sperm would mean death for the zygote.

Now, it seems, nature hurls the sexes at each other. The news out of laboratories tells of wars between the sex cells, the battlefield being the female reproductive tract, the weapons physical or chemical. In species where males compete for females or females mate with males in quick succession, there are various strategies of sabotage. Chimpanzees churn out huge numbers of sperm to swamp the competition. Certain insects evict enemy sperm before depositing their own. One theory, unwincingly described as the Kamikaze Sperm Hypothesis, proposes that in these polygamous species, sperm evolved specific features for combat with the sperm of other males. Even the two hundred million or so sperm in a man's ejaculate, says the reproductive biologist Robin Baker, are not necessarily born alike. To some, nature has added a little violence of direction. There are relatively few actual egg-getters; the rest are "blockers," which thwart the advance of sperm from rival males, and some with a "seek-and-destroy" mission. However, when scientists tested the idea in humans, mixing sperm samples from more than a dozen men, they found no evidence of any such spermatic battles.

Even so, this kind of warfare may be at work in other species. Male fruit flies lace their seminal fluid with up to sixty proteins designed to boost their chances of beating the sperm of other males. Some of the proteins dampen the female's sexual appetite and make her quickly lay more eggs; others kill off the sperm of rival flies, poisoning the female in the process.

Females, in turn, have their own strategies for keeping at bay more than the desired single sperm. Female fruit flies have a bag of chemical tricks, proteins that kill off or counteract the effects of the male proteins. In mammals, when a sperm enters an egg's membrane, thousands of granules sent out by the membrane change its structure, releasing sperm already bound there and blocking the entry of others.

That human fertilization is chancy I know from experience. During the two and a half years it took to conceive my first child, I resisted the temptation to reckon the math of it, but I couldn't help investigating all the possible obstacles: for my husband, the past hazards of mumps, measles, whooping cough, diphtheria; the present risks of imbibing substances with the power to lower sperm count, such as alcohol, tea, coffee; the conceivable blockage of seminal vesicles from sexually transmitted diseases; the risk of a glitch in the Y chromosome that prevents sperm production. And for me, the off chance of lurking fibroids, tipped uterus, hormonal imbalance, aging ovaries, allergy to sperm.

Given the sea of contrarieties and possible blockades, the surprise is not that some sperm and egg make shipwreck but that any come to port. They do, of course, and recent news suggests that they overcome their differences and contrive to join, like any green couple, by means of subtle, sophisticated, nuanced talk.

Even before they meet, sex cells communicate by means of a "come-hither" signal from the egg, a chemical guide for the sperm — or lure, depending on your perspective — as it weaves its way from vagina to fallopian tube. Contrary to the popular belief that eggs are sleeping beauties awaiting a sperm's kiss, they actually get sperm going. The signal may well be a holdover from a time when our ancestors fertilized their eggs outside the body in the open sea, where the trick was to find mutually compatible gametes and to make sure starfish sperm didn't fertilize urchin eggs. Put a swarm of sea urchin sperm in a tank of seawater, and they'll swim about in wide, lazy circles. But squirt a drop of fluid with this attractive chemical in the middle of the tank, and the sperm will dart toward the drop. Like hounds on the trail of a fox, they'll swim up a concentration gradient of this compound until they reach the egg, then congregate there, one imagines, howling for entry.

Though human eggs and sperm are not free swimmers in the open sea, they apparently rely on similar signals. One whiff of the fluid from a human ovarian follicle — a cell cluster that surrounds developing eggs — and human sperm will speed toward the source, sniffing out the signals with olfactory receptors identical with those we use to detect aromas. Only a fraction of the sperm cells respond to the signal; the rest are repelled or deactivated by the same factor, suggesting that this molecule, whatever it is, may be a way for the egg to select one "competent" sperm.

Once contact is made, more talk ensues, a swap of signals to ensure that egg and sperm recognize each other as members of the same species. In most organisms, sex cells are adept at spotting their kind. For good reason. Embryos resulting from cross-species fertilization usually fail to develop normally. In plants, the stigma, the receptive female part, discriminates among its suitors by conversing with the cell wall of the pollen grains; it will hold on tightly to pollen of the same species but release grains from a different species.

The sperm of sea urchins spot their species' eggs with the help of bindin, a protein lodged in their membranes, which matches a complementary receptor on the egg surface. Humans and other mammals have a similar set of proteins that effectively tether a wriggling sperm to the surface of the egg in thousands of places. In mammals, from mice to horses, the pattern of these proteins — their distribution and arrangement on the egg surface — is remarkably similar. But the makeup of the individual proteins is species-specific. These attachment proteins perform the final check to make sure egg and sperm are a suitable match. As such, one would think they wouldn't vary much within a species. But scientists have discovered that the bindin of different sperm in one species of sea urchin do come in distinct flavors. Sperm with one variety of bindin are better at fertilizing eggs than those with another. The bindin may be evolving to keep pace with the receptors on the egg, which are shifting slightly to ward off fertilization by more than one sperm. Scientists speculate that this kind of matching game between egg and sperm proteins may be a first step on the path to the creation of a separate species.

I once compiled a list of probabilities; among the odds I happened on were these:

> that I might crush my finger with a hammer: 1 in 3,000
> that an asteroid might collide with our planet in the next millennium: 1 in 100
> that a mutation in my DNA would be favorable: 1 in 50,000
> that my pregnancy would come to term: 1 in 3
> that my baby would have a birth defect: 1 in 20
> that my baby would be what it is, the product of an accidental joining of one sperm (in millions) and one egg (in hundreds): 1 in 3 billion

How marvelous that my child's precious existence would come about through such chancy biologic success. How remarkable that the possibility of each of us is such a slim margin against great statistical improbability.

When a single lucky sperm finally, improbably, penetrates an egg, an explosive release of calcium stimulates the egg, and its insides begin to stir and jiggle, driving the sperm's nucleus toward its own until the two meet. Twelve hours later, the chromosomes mix to make the next generation.

"It is perhaps one of the abiding mysteries of genetics that two genomes can ever come together in one cell, reside harmoniously together, and then peaceably disperse (i.e., go through Mendelian segregation)," writes Laurence Hurst, an evolutionary biologist. But this last-minute mixing of genes is only the final act in a long drama of gene commingling. By the time sex cells meet, they're already carrying a unique blend of genes, created when the cells were first made. By the time we are old enough for sex, the quip goes, our germ cells have already had half the sex they are ever going to have.

Here's where nature breeds the possibilities. During the specialized form of cell division for making sperm and eggs — known as meiosis — the chromosomes that have been inherited from mother and father double. Then the new chromosome pairs find each other and line up, with each chromosome from an individual's father joining with its counterpart from the mother. The pairs embrace and swap pieces of DNA. It's a glitch in this stage that often results in a gene's being duplicated and added to the batch. This, too, is the moment when deleterious mutations that might cause genetic disease are removed, one of the great advantages of sex. (The human genome is plagued with a high rate of mutation, say the latest studies, a hundred per generation, of which several may be harmful. If it weren't for sex, with its ability to foil outlaw genes, notes the biologist Adam Eyre-Walker, our species and many others would probably have vanished.) Once the chromosomes have pooled, the cell divides; into each newly created egg or sperm goes one copy of a completely new blended chromosome (sometimes with a duplicated gene slipped into the mix), eventually to be passed along to offspring. Because of the two-step blending of chromosomes, the number of possible gene combinations in any of us is mind-boggling.

Even after the marriage of sperm and egg and the joining of their chromosomes, genetic conflict may take place, a race between maternal and paternal genes within an embryo. Mendel taught us to expect the equal participation of the genes we inherit from each parent. But in a small group of mammalian genes, only the mother's or father's copy of a gene is expressed; the other is gagged. Many of these so-called imprinted genes govern fetal growth. David Haig of Harvard University has proposed that imprinting arose because of the conflicting interests of males and females. If the goal is to make as many progeny as possible, the mother would do best to hold back on the resources with which she supports any one litter, saving up for future litters. The father, on the other hand, has no genetic investment in the mother's future litters, so he wants this one to have the biggest possible babies. The father's genes try to muzzle the mother's to promote growth; the mother's genes aim to gag the father's to limit growth. It's an early version of parental tug-of-war.

In the end, though, the self risks conflict, even chaos, to merge with another. Think of the sperm, blind and headlong, and the egg, open-eyed and active, both burning with strategy. Their union is really nothing but an all-out commitment to creating, against all odds, a glorious new knit of identity, a fearful, wonderful blend of ancestors, to be found in the angle of a cheekbone, in verbal facility, in a wry sense of humor.

Here's why I vote for sex over cloning. It's the perfect way to fiddle with possibility, to get a fresh mix, a new song among songs. Thanks to its invitation, its peculiar mingling of risk and possibility, splitting and binding, conflict and cooperation — which most of Earth's creatures seem eagerly to embrace — variety has come to be one of life's enduring familial traits, its transfiguring beauty.

5

CELLE FANTASTYK

So MUCH for the minute, mixed-up grant of our inherited wealth. Now to the fertilized cell, the mansion of our birth.

Once, in a schoolroom suffused with sunlight and the music of Bach, I saw the fertilized egg of a frog magnified for cool study under a microscope. The Bach was an indication of my biology teacher's belief that the patterns of music, the ascending and descending tones, the musical fourths and fifths, could help her students' concentration and memory. The egg was a simple sphere soon to be split by a single groove, we were told, setting out what would become the left and right of the animal. We were asked to imagine what might happen next, the second split at right angles to the first; then, the horizontal slice, girdling the egg near its north pole; later, this simple grid of life broken into myriad cells, ever finer, ever more specialized — an extraordinary process, our teacher reminded us, given that every cell in that developing embryo carries exactly the same genetic information in its DNA, the same score. Somehow, the cells of the developing tadpole knew when to play their portion of the score, sorting themselves into top or bottom, left or right, front, middle, or rear, then into heart, liver, eye, chest, until days later we might see the awkward head take shape at one end, tail at the other — the beginning of clamorous, croaking frog.

When I first conceived, I remembered that frog egg, imagined my zygote's patterning conjoined with Bach's mathematically precise phrases: the single cell cleaving into two cells knobbed with possibility, then four, then eight, and so on, the myriad little progeny eventually diversifying into trillions of individual cells of splendid variety, all toiling together for a single being.

Chaucer wrote of the *celle fantastyk* of the brain, but he meant the word in the Latin sense, a small room or cubicle, like a monk's. As Henry Harris has written in his excellent history, *The Birth of the Cell*, before the advent of the microscope, early thinkers believed that the elementary components of the body were fibers — not surprising, given the visibly threadlike nature of the body's vessels, tendons, muscles. The fibers came in three kinds, said Galen in the second century, straight, transverse, and oblique. No, two, argued Giorgio Baglivi fifteen hundred years later, the fleshy and the membranous. Three, asserted the Swiss anatomist Albrecht von Haller in 1757: the connective *(tela cellulosa)*, the irritable *(fibra muscularis)*, the sensible *(fibra nervosa)*. A prominent French physician counted twenty-one varieties. In any case, stated Haller, "a fiber is for a physiologist what the line is for the geometer, that out of which all other figures are constructed." Haller believed the fibers were made of long strings of atoms stuck together with gluten and molded by pressure. These fibers formed all the parts of the body — skin in a tube, muscles in bundles, organs like pipes.

"Cell" was coined as a scientific term in the seventeenth century by Robert Hooke to describe each of the chambers he discerned through his microscope in insect parts, bird feathers, fish scales. Hooke was a polymath, like Goethe a century later. A scientist who studied painting, he made a name for himself in physics by fathoming the behavior of light, the theory of gravity, and the action of springs. Holed up in his lab amid the ravages of the Great Plague, Hooke the biologist placed a sliver of cork under his microscope and saw "an infinite company of small Boxes or Bladders of Air," which explained the lightness of cork; it was "the lightness of froth, an empty Honeycomb Wool, a Spunge, a Pumice-stone." There were more than a thousand such pores or cells in an inch, more than a billion in a cubic inch. These cells, stupefying in their numbers, were not peculiar to cork, Hooke discovered, but were present in "the pith of an Elder, or almost any other Tree, the inner pulp or pith of the Cany, hollow stalks of several other Vegetables: as of Fennel, Carrets, Daucus, Burdocks, Teasels, Fearn, some kinds of Reeds."

Hooke saw not cells themselves but their skeletons, the walls that remained after the living cells inside them had dried. Neither he nor his peers imagined that these walled spaces were anything but microscopic voids, delicate channels or pipes "through which the Succus nutritius, or natural juices of the Vegetables, are conveyed," he surmised.

The cellular nature of plant tissue was discovered in 1671, simultaneously and independently, by both Marcello Malpighi, an Italian biologist, and Nehemiah Grew, an English physician. Grew described plant tissues as bladders clustered together, and he drew them beautifully but believed they were themselves woven from fibers.

Not long after, Leeuwenhoek wrote to Hooke to tell him that he had seen in the blood of a ray "flat, oval particles, thickish floating in a crystalline water," but neither scientist considered the particles even remotely related to animal tissue. It was the French, a century later, who finally realized that animals, too, were made of cells. On examining connective tissue in a human thorax, a dog's paw, a cockerel, a frog, and the abdominal cavity of a carp, the naturalist Henri Milne-Edwards found everywhere "elementary globules, in their shape and in their diameter, similar to those that are to be seen floating in pus, in milk, and so on." He went on to find the same in muscle, skin, and nervous tissue, and in the gray and white substance of the brain.

The fundamental likeness of plant and animal cells was noted in 1835 by a young medical student studying mammalian tooth enamel. But the matter was settled, and the idea elevated a few years later — according to one version of the tale — in a supper conversation between a botanist and a zoologist.

Matthias Jacob Schleiden, an expert in the microscopic anatomy of plants, was expounding on the important role of the nucleus in the development of plant cells. Theodor Schwann, who studied the nerves in the tails of frog larvae, was listening intently. "I at once recalled having seen a similar organ in the cells of the notochord," he later wrote. When the two scientists compared notes in Schwann's lab, they found striking similarities in the nuclei of animal and plant cells and declared, "We have thrown down a grand barrier of separation."

Schwann went on to study bird eggs, horny hoof tissue from a pig embryo, sparrow feathers, the eyes of a young pike, the cartilage of rays, innumerable samples of tendons, muscles, nerves, glands, and the fingernails of a newborn baby, concluding that all animal tissue, as well as plant tissue, was made of microscopic cells, each cell a minute living entity of its own.

"Nature never joins the molecules together in a fiber, tube, etc., but always first fashions a cell." He likened an animal to a hive of bees, an assemblage of independent entities. Schleiden agreed. "Every cell thus leads a double life," he wrote, "one that is entirely self-sufficient, and pertains only to its own development, and another, indirect, insofar as the cell has become an integral part of a plant."

Soon every living thing revealed itself as a cell or collection of cells stuffed with what appeared to be a living jelly, or protoplasm, a granular suspension endowed with the ability to move, grow, excrete, and give rise to everything in the body, and animated by a ghostly "vital force," like the breath of God. "Life is one," wrote the nineteenth-century French biologist René Joachim Henri Dutrochet. "The differences shown by its various phenomena, in all things that are alive, are not fundamental differences: if these phenomena are tracked down to their origins, the differences are seen to disappear, and an admirable uniformity of plan is revealed."

Once, finding himself bleeding profusely after a bad fall, the naturalist Loren Eiseley apologized to his doomed blood cells: "Oh, don't go. I'm sorry." The words were spoken to no one, he wrote, but addressed to all the "crawling, living, independent" entities that had been part of him, and now, through his "folly and lack of care, were dying like beached fish on the hot pavement."

I'm not sure I could summon such compassion for my individual cells, not sure I could think of them as creatures with a wit and wisdom of their own. Just as a single molecule of H_2O doesn't make water, a single cell from the brain of a mouse sliding over its petri dish contains no thought; only when the cell links up with millions of others in a network to shuttle electrochemical impulses does thought emerge.

Still, Eiseley was on the right track.

One afternoon not long ago I watched a film of a solitary creature, like a beautiful little squid, probing its chamber. I have long admired squids and other cephalopods. Strong, agile, over short distances they're probably the swiftest, stealthiest creatures of the sea. A cephalopod's head — *cephalo* — connects with its feet — *pods* — forgoing any sort of torso. Its beauty is the radial beauty of the starfish. Its skin is shifty, capable of changing color faster than a chameleon's, flushing red when the creature is excited, bleaching white when it's bored or frightened. Cephalopods are old, born sometime in the late Cambrian age, and have slipped through many shapes, into shells as nautiluses, and out again as squids and octopuses, trading hard protection for camouflage, agility, complex behavior, and evolution of elaborate brains. (Here was an evolutionary gamble from which we've benefited. The giant neurons of squids, a hundred times larger than those of a human, have contributed immensely to knowledge of our own neuronal function.) They are the wisest of the mollusks, solitary for the most part, but also curious and exploratory. Octopuses given a floating pill bottle will jet water at the bottle so that it circles their tank, purposeless behavior that looks a lot like play. And why not? Wolves, hyenas, bats, anteaters, even sea turtles engage in play — "aimless" activity that may stimulate the formation of synaptic connections between brain cells, help muscle tissue to mature, provide practice for hunting, mating, and rearing young. People who work in laboratories with *Octopus vulgaris* say that it's difficult to study the mollusk objectively, because it is always studying you. I believe this, having spent time staring through the glass at an octopus drifting in the gloom of an aquarium tank, a Hindu god with slotted, hypnotizing eyes.

In the film that afternoon, the little squidlike creature in the chamber thrust forward a slender foot, as if to test the possibility of moving in one direction. Then it pulled back the foot slowly, paused for a moment, and sprouted a new foot. This thin, twitching new limb poked out in a different direction, again searching like a tongue or antenna.

Neither time nor scale was real in the film. It was time-lapse videomicroscopy, with the small made large beneath a high-power lens, and hours squeezed into seconds — the same kind of compressed-time photography that makes clouds boil across the sky and the tendrils of a growing vine snake around the trellis. Accelerated or not, the movement of the little beast looked willful, no mere amoebic slithering or crawling about, but a delicate, precise, silvery motion invested with a cephalopod intelligence.

The film was the creation of Diane Hoffman-Kim, a cell biologist, then at Harvard. Her star was no mollusk but the amoebalike growth cone at the tip of an axon, a slender fiber that extends from a nerve cell — this one isolated from the spinal cord of a chick embryo. The "feet" were filopodia, Hoffman-Kim told me, dynamic little extensions of the cell membrane thrusting out from the growth cone, the leading edge of a growing axon, which the cell uses to sniff out connections with other nerve cells.

Hoffman-Kim was studying the way cells find one another to create the netted wiring of the brain, not just during embryonic development but all through life as the body learns new skills. She was interested in the molecules outside the cell that steer the tentative axon to its quarry, telling it where to go as it tunnels through tissues, in this direction or that. Once it finds exactly the right cell to hook up with, a link is formed. Somehow the cell initiates the connection and is itself altered by its accomplishment. If a skill is used again and again, the link hardens; if it's used only once, the filopodia eventually shrink back, triggering a change in the brain cell, a thinning of the grip that may explain forgetting.

I was fascinated by the disembodied cell itself, which looked for all the world like its own creature, if not a cephalopod then a protozoan from Leeuwenhoek's scummy pond. It certainly acted independently, crawling about its culture dish as if it were bottom-feeding or seeking out its brethren. Under the right conditions, Hoffman-Kim explained, any cell can enjoy a kind of free existence no less sophisticated than that of a single-celled creature rioting in a roadside pool. Neurons, in particular, have something like a mind of their own.

On the scale of biological complexity, the cell is said to occupy a spot roughly midway between the molecule and the organism, between the microcosmic world of genes and proteins and the visible world of the squid. My cells may have subsumed their individual identity by joining a larger community, but each still has its independent existence and contains a similar, if simpler, version of the mystery hidden in my head. Each is an immensely competent being, holding my whole genome, capable of subtle movement, rhythm, sophisticated talk with the outside world, adept at sucking energy out of life, and — when sufficiently full of zip — splitting, doubling, and redoubling to make the world a fecund place.

Who's to say that each does not enjoy its own life?

Cells of yeast, fly, botanist, all make their living in nearly the same way. Overlap in the genomes of yeast and worms suggests that a core group of genes, about three thousand, are critical to the basic workings of a cell. The genes that code for proteins involved in DNA replication or protein synthesis, for instance, are virtually the same in yeast, worms, higher species. Likewise, the enzymes that orchestrate chemical reactions in cells are almost everywhere identical. So are certain proteins that allow for movement. Most cells creep and shift shape by means of microtubules, protein filaments that radiate throughout the cell, allowing it to crawl from here to there, to squeeze through vessels, to come together to form tissues without losing its integrity.

Or consider growth. A cell doubles by doubling everything it is made of, every part, every kind of molecule, even the amount of water it contains. Then it divides in an orderly sequence called the cell cycle. (The time it takes to double depends on the number of genes in a given genome; for bacteria, it's about twenty minutes; for a human cell, with a thousandfold more DNA, about twenty-four hours.) The genes that control the cell cycle in my body are the same as those that dictate the doubling of cells in my fungal cousins. In fact, human versions of the genes will function perfectly well in the cells of baker's yeast — good evidence of an early common ancestor with some cunning energy of invention.

Beginnings are apt to be shadowy, and so it is with the first cell, born perhaps more than 3.5 billion years ago in a triumph of organization. One crucial invention was a membrane, a barrier to hold inside in and outside out, a microscopic wall against the world to protect the necessary molecules and their chemical reactions, a fortification of being against not-being. The first cell may have resembled a modern bacterium or archaean — small, simple, with no nucleus, a "prokaryotic" cell (from the Greek *pro,* "before," and *karyon,* "nut"). Such cells were the sole occupants of Earth for much of its history, flourishing in pools, hot vents, and rock crannies, blossoming across the planet, transforming its atmosphere like little alchemists, learning to hoard energy from the sun, perfecting their molecular wisdom, inventing sex and memory and communication.

It took another billion years or so for our kind of "eukaryotic" cell to emerge (*eu* from the Greek for "good"), with its little nut of a nucleus and other complicated internal parts, many of them enfolded within their own membranes. That first eukaryote was likely born not as a single entity, as we once thought, but as a primordial merging of separate organisms.

When I look at a liver cell through a microscope, I see hundreds of lozenge-shaped bodies. Time-lapse videomicroscopy would show them shifting about, changing shape, even fusing and separating. These are the mitochondria, the organelles that nurse cells so carefully select and deliver to the almighty egg. Mitochondria make adenosine triphosphate, or ATP, the magic molecule that serves as the principal carrier of energy in all animal cells, providing the motive power for nearly every activity. Lay all the mitochondria in your body end to end, and they would circle the Earth two thousand times. Without these prominent little generators, I could not flex my toe, wag my tongue, jog my memory.

Mitochondria are about the size of bacteria, and are like them in other ways, harboring their own set of genes (tucked into small loops of DNA like the loops found in *Escherichia coli*), reproducing on their own, independently of the cell, and doing so just as bacteria do, by fission. Their membranes, too, closely resemble those of bacteria. So do their ribosomes. They look for all the world like little microbial cells within our cells, separate creatures capable of leading an independent existence — and there's every reason to believe they once did. The evidence is strong that cells of fungi, protists, plants, and animals were pieced together by the joining of two or more ancestral cells about a billion and a half years ago, and that our DNA carries genes gleaned from these invaders.

I like this notion of newness arising not just from a chance mutation or the random jumping of a gene, but from a sort of joint venture, a fizzling of the line between two beings. So, too, I relish the idea that from the descendant of an ancient form of bacterium, some old celle fantastyk, I draw the juice that enables me to perceive, remember, mull over the future, and speculate on the possibility of cephalopod play.

Fungi have four types of cells. Sponges and jellyfish have nine to twelve. Duckweed about twice as many. Fairy flies have something like fifty-five. My body has hundreds of varieties with an exaltation of names:

mast, mucous, hair
stem, spindle, squamous, synovial, serosal
Merkel, Boettcher, Clara, Claudius
Schwann, Sertoli, Purkinje

isolated goblet
basal, brush border
fenestrated
fat
germ
killer
nurse

Nurse differs from germ, Clara from Claudius, Schwann from goblet, because each type of cell "decides" to express its specialized set of genes at some point during development and thereby produce its specialized proteins. But underlying every cell is the same genome. The genes for hemoglobin lie quietly in skin cells; the genes for keratin slumber in the cells of the brain.

Unlike single-celled organisms, creatures made of many cells need a genetic control system that will tell different genes to blink on or off in different cell types as development unfolds to make a body plan. How does a young cell choose its destiny as the fertilized egg divides, divides again, and generates a multiplicity of distinct cell types? What makes a cell decide what to be and when to be it?

To fathom this is the dream of H. Robert Horvitz, a developmental biologist at the Massachusetts Institute of Technology. Horvitz studies the lineage of individual cells in the developing embryo — their parentage, siblings, daughters, distant descendants — and the reasons for their being. He has great affection for his experimental subject, a lithe little nematode worm known as *Caenorhabditis elegans*, or *C. elegans*. Jonathan Edwards called the worm "a little, wretched despicable creature . . . a mere nothing, and less than nothing; a vile insect that has risen up in contempt against the majesty of Heaven and earth." Horvitz calls his nematode a "tiny person in disguise."

Though *C. elegans* is small (a millimeter in length and made of fewer than a thousand cells; ten thousand worms can fit in a petri dish, leaving inches to spare), the nematode nevertheless has perfect working systems that share many features with those of humans: muscles, intestines, sexual organs, and a nervous system that includes a simple brain. So transparent is the worm at all stages of development that one can easily see, under a microscope, individual cells going about their business — dividing, migrating, dying. By tracing the fate of each cell as the worm transforms from egg to full-size being, biologists have created a complete family tree of nematode cell lineage.

I think of developing cells as balls rolling down branches of this gently undulating family tree. As time passes and the cells divide, the branches fork. At every fork, a cell must decide whether to be like its mother cell or be different. Which branch a cell follows depends on the signals it receives at the fork, either from within or from without. Those signals determine which genes are switched on or off inside it, which proteins it will produce, how it will behave, what sort of cell it will become. It's a string of little wows that builds to the dramatic moment when the worm moves and acts as one organism.

Bob Horvitz seeks to plumb what factors shape a single cell's fate, which signaling molecules may push the genetic switches, turning genes on and off as the cell develops, sending it down one branch or another. He has found that the genes shaping the fate of cells in *C. elegans* are strikingly similar to genes in other organisms, including humans. One gene that dictates the decisions of cells making up the vulva of the worm belongs to the Ras gene family, also active in shaping the eyes of fruit flies. Ras genes have turned up in humans, too. Proteins made by mutant human Ras genes disrupt the normal controls on cell growth and differentiation, sometimes causing cancer. (As Nietzsche said, "You have made your way from worm to man, and much in you is still worm.")

In nematodes, the identity of any one cell may be determined by its ancestry — what it's born with — or by its surroundings, the signals it receives from its neighbors. The very first division of the fertilized egg in the nematode creates two different types of cells, two sisters, thanks to special molecules inherited from the mother worm. The cells born in the second round of division are shaped by signals from neighboring cells.

As our own cells divide, they follow this second, social method; they listen to other cells. Here is a startling secret of development. As our cells roll down the branches of lineage, they converse, conspire, and collaborate, as loud and gregarious as a flock of gulls, telling one another what to be and when to be it. The chatter picks up when they're in a crowd. So potent are the messages that cells remember their assigned identities long after the actual cues and signals have disappeared. The messages may be brief, but they're as vital and precise as the eighth notes scratched on a composer's score. If a barrier as frail as a sheet of cellophane is placed between interacting layers of an embryonic disk, the whispers go unheeded, and normal development ceases.

It may be that the need for this cell-to-cell conversation explains the long pause before the appearance of the first multicellular creature. It took time for cells to learn to signal one another in development, to coordinate their lives for the benefit of some new, whole conglomerate being.

I like to imagine the thin, urgent whispering as the human fetus develops, the notes flying as the cells gather into specialized groups, the groups into layers, the layers into sheets and tubes, sliding into place amid the jumble, forming limb where limb should be, heart where heart belongs, until there it is, emerging all of a piece from nose to tail, consistent with the score: warm blood, no gills, brain, spinal cord, beating heart, scant patch of hair.

6

GENERATION

WHEN I WAS PREGNANT with my first child, I was shocked by the idea that it required no thought at all for me to sculpt a whole other person. That it was somehow built into my species — all species — to generate copies of itself, not identical, of course, but remarkably alike in the spectrum of life.

I wanted perfection. I imagined the baby's growth as a slow-motion, silent explosion inside me, neat and orderly, like a chain reaction. I willed one perfect cell to form, and another, a million perfect cells, body shape uncurling or building up, beginning with bones — white weightless wires, naked at first, then growing the thinnest silk of flesh. Cell after cell, the bones would gain heft, the muscles thicken to fill the spaces of the network.

But at night I worried about what I held. The genetic burst that delivers the knack for numbers or musical talent, the stray taste for chemistry, the hand for drafting or bones for dance, can as easily blight limbs or punch a hole in the heart. That my sister Beckie had been born microcephalic — head too small, body twisted, intelligence trapped in infancy — blunted the expectant buzz of maternity. Beckie had slipped from dark into light, a primal piece of life, shaped and unshaped. Over the years I watched her helplessness drive my mother to resignation, at times to despair. My mother's grief belonged to the phylum of fatigue, the problem of how to get through the days, seasons, years, with a child who would never feed herself, never talk or walk without help, never know the childish pleasure of naming dinosaurs or birds.

It was little comfort knowing that the root of my sister's deformity lay not in ancestry but in some other secret, perhaps a first-trimester virus, a wayward scrap of DNA that found its way into my mother's blood and unhinged her baby's whole development. I was told by relatives how unlikely it would be for a family to suffer twice in this way, as if family misfortune could be drawn off in one member, the others somehow proportionally relieved. Though I knew my chances of having a child like Beckie were no greater than those of any other mother, I did not hold much stock in odds.

My imagination played in a thousand dark ways. The horrors would gather in me as a migraine does, a flicker of imminence before I knew it was there. The baby would grow in the S-curve of a serpent, a sheathed continuity of form without limbs. Or it would grow from one end only, lower limbs and torso withered, birdlike or less, enormous swollen head pushing into my lungs until I gasped for breath. I imagined profound deformity drifting there in the dark, a clot of furled flesh. I dreamed and dreamed again of washing my blotted baby — no, sister, no, baby — down the drain, methodically, every last bit, then realizing in heart-stopping horror what I had done. In and out of my dreams floated the grotesque image of an ant lion, that insect with a name like an oxymoron or an odd chimera from my childhood bestiary. In its adult form *Myrmeleon immaculatus* is lovely, long-bodied like a damselfly, with a pair of ethereal wings. Its larva, the ant lion, is a queer, wedge-shaped thing with long, bristling sicklelike jaws and a nasty habit of sucking the juice from its prey. Metamorphosis is the law of the universe. If an insect can contain both winged elegance and hideous nymph, why couldn't this germ of a baby?

I taunted myself with medieval ideas that pregnant women transmit marks of their fancies to the bodies of the children in their wombs, old notions that my dark thoughts themselves might unfavorably alter the form of the fetus. As antidote I thought of dancers and athletes. I summoned the nudes of Leonardo and Michelangelo, bodies in every conceivable attitude exploding with perfect proportion, form, and grace. Still, it took a wing stroke of will to loose the nightmares and find peaceful sleep.

In the immense animal pressure of labor, my fears evaporated, and when my daughter arrived, sweet and sound as a nut, I accepted her perfection as given.

During those nine months on guard came a day that was significant in the history of biology, because it was the day on which the discovery of a new creature was announced: a little organism never before seen, living on the lips of Norway lobsters. The new creature was called *Symbion pandora*: *Symbion* for its intimate life with the lobster; *pandora* for the bizarre box-within-a-box form it takes during one of its life stages. Less than a millimeter in length, it looks like a Lilliputian sac, its foot a sticky disk, its mouth a splendid circle of whirling cilia, with an anus just adjacent. So radically different in form is *Symbion pandora* from any other animal that science created a special phylum for it: Cycliophora.

The unearthing of a new species is not a rare event, of course. New kinds of insects, mollusks, even mammals are turning up everywhere we look. People living on the island of Panay in the central Philippines found a new species of rodent, the Panay cloudrunner, an agile squirrel-like, nocturnal creature with the shrill cry of an insect. In the Annamite Mountains bordering Laos, a new species of striped rabbit was discovered, and from the underbrush along the backbone of the Ecuadorean Andes, a hitherto unknown bird, plump, long-legged, a kind of antpitta with a haunting, hollow, hiccuping call. Scientists exploring the sea's dim recesses lately clapped eyes on a giant squid, not a new species, but an animal so elusive as to disappear for centuries and reappear only in the guise of a monstrous tentacle washed ashore. *Architeuthis,* with eyes the size of dinner plates, looks like something etched in the sea foam on an ancient map. News of smaller but no less interesting quarry came not long ago when marine biologists plumbing a patch of deep-sea ooze off the coast of New Jersey hauled up buckets of worms, jellyfish, anemones, corals, snails — close to eight hundred species, more than half of them new to science.

But the discovery of a whole new phylum! The animal kingdom has only thirty-five basic phyla, each defined by the distinctive body plan of its members. Those hundreds of ooze-dwelling New Jersey creatures, stunning though they were, all fell neatly into known phyla with familiar body plans: nematodes, annelids, sipunculans. *Symbion pandora,* on the other hand, commanded its own singular category, a taxonomic Oscar. And it turned up not in a remote mountain spur or the abyssal depths of an oceanic trench or in the recesses of the last surviving rainforests but right under our noses.

In Linnaeus's day, the number of recognized species was about twelve thousand; today the number of named plants and animals is roughly two million. Into this vast, sprawling scheme has crept the fact of *Symbion*. The discovery of the little oddity from the lips of lobsters, which demands not just a place on an obscure branch in the tree of life but a whole new bough, pleases me immensely — more so than, say, the sighting of a new planet. Animals exert a special hold on the human mind. Studies of the brain suggest that special areas in our gray matter are highly sensitive to certain categories, chief among them animals. People who have suffered damage to a region of tissue at the back of the brain fail to recognize skunks, tigers, cats. Those with damage in a different patch can recognize the animals but can't recall their names. The theory goes that the two patches communicate. When we see a skunk, the cluster of brain cells specializing in animals may tweak the neurons in the brain region where the word "skunk" is retrieved. I like this notion that animals possess substantial niches in our cerebrum, one for themselves, one for the names we give them.

Symbion pandora. With a name like yours, you might be any shape, almost. *Symbion* looks for all the world like a toe loosed from the old life. It reminds me of those dreamlike creatures uncovered in the fossil beds of the Burgess Shale in the Canadian Rockies, eight thousand feet above sea level: five-eyed *Opabinia; Anomalocaris,* with a mouth like a circular nutcracker; *Hallucigenia,* its bulbous head and body suited to its name; flattened *Amiskwia;* elegant, segmented *Pikaia.* In these limestone remains of an ancient sea, life that turned to stone six hundred million years ago, is a fantastic diversity of forms: insects, earthworms, mollusks, which seem to have come into the world all at once. The generous bloom of new phyla in the Cambrian period has no equal, not even in the times following the mass extinction at the end of the Permian period, when 95 percent of all marine species disappeared, and there must have been a lot of empty niches to fill.

Just what sparked the explosion is still a matter of debate: a rise in atmospheric oxygen, shifts in ocean currents, the opening of warm, shallow seas, or, possibly, polar wander — a listing of the Earth, which sent continents soaring across its surface, turning topsy-turvy entire ecosystems and opening the door for new species. One of my favorite theories suggests that the Cambrian bloom had distinctly humble roots. At some point more than half a billion years ago, when organisms evolved a one-way digestive tract that expelled fecal pellets, the ocean was chemically transformed in a global rain of feces, which opened up the seas' depths to colonization by myriad organisms.

Whatever the reason, the Cambrian era produced a menagerie of forms, all the major body plans of today, even the weird plan of *Symbion pandora*. (Scientists have in their hands the fossil of an arthropod from the Cambrian period that looks as if it may have hosted like-bodied ancestors of *Symbion*.)

The world is full of creatures if not new in shape, then strange. I think of the ant lion, the hammerhead shark, the pangolin, with its serpentine tail bound to a body like an artichoke. Even within a single phylum, the possibility of form seems infinite in all directions. Vertebrates alone possess a profusion of shapes — hagfish, platypus, spiny anteater, slimy sculpin, star-nosed mole, with its radial snout of twenty-two fleshy rays that grope through dark soil for edible morsels. Arthropods, with their hundreds of thousands of species in marvelous shapes, put the rest of life in the shade. As Jorge Luis Borges wrote, "The zoology of dreams is far poorer than the zoology of the Maker." And when you consider that what lives today represents only a tenth of all known forms of life, when you take into account all that have vanished — animals with myriad wings, teeth, horns, and tails sculpted by natural selection — the panoply of living forms seems almost unimaginable.

And yet the basic body plans for all were present in the Burgess Shale. Not a single new one has appeared during the last half billion years, and some have gone the way of *Opabinia*, into evolution's dustbin. In fact, as Stephen Jay Gould has written, a cardinal feature of modern life is stereotypy, the cramming of millions of animal species into just a few basic anatomical plans.

What I had been wishing for during those worrying days of pregnancy, beyond full-size brain and sufficient limbs, was bilateral symmetry, one of those ancient anatomical plans, the body tidily divided into like halves along an axis from head to toe. There are other forms of architecture, of course. The radial elegance of the medusa bell, the cephalopod, the starfish with arms extending in all directions. Or the spiral, the circle unwound and set free: the sinistral spiral of a honeysuckle vine, the horns of wild sheep, the cochlea of an ear, the human umbilical cord, and, especially, the shells of certain mollusks, expanding geometric figures that sweep out in coiled edifices of exquisite beauty. Radial symmetry is an ancient form of organization, one eminently sensible for the sessile creature, allowing it to feed in a circle and respond to danger from all directions. But the human notion of fine form is bound up with the bilateral, the symmetry of left and right, the pairing of limbs for efficient locomotion and for beauty, the balanced hips and legs, the eyes equidistant from the nose, the lips curling out from a precise central axis on the face.

The majority of animals are bilaterally symmetrical, including flatworms — small brown fleshy blotches that look only one step up the evolutionary ladder from mud — as well as the wonders of the Burgess Shale, even *Symbion pandora*. We are not alone in relishing the trait. Many animals see the beauty of bilateral symmetry and prefer it: apes, dolphins, birds, even bees. Bucks with the biggest, most symmetrical antlers boast the biggest harems. Female swallows select mates with perfectly symmetrical forked tails. Japanese scorpion flies prefer the scent of a male with a symmetrical body. Lately, we've heard the somewhat disturbing news that symmetry may reflect the fundamental health of an individual, the strength of immune cells, the robustness of genes. One study showed a modest link between bodily asymmetry and lowered IQ; another went so far as to suggest that men with asymmetrical hands had low sperm counts and poor sperm motility. A developing organism stressed by poor nutrition, disease, inbreeding, or almost any genetic defect will exhibit some visible asymmetry. Perfect symmetry, in turn, may signal internal well-being. Despite my mother's arguments to the contrary, it seems, beauty is no weak guarantee.

And yet, in the throes of gestation I had also been wishing for asymmetry. Our insides hardly reflect a looking-glass world. Beneath the skin, symmetry vanishes. Heart, liver, spleen, pancreas, gut, all lie to one side of the body's midline, and those organs themselves are largely lopsided, the intestines tortuously looped and coiled to fit neatly in a small cavity, the heart divided into four irregular chambers and netted with a maze of curving blood vessels that send blood through it in swirling patterns. The inner body's extravagant asymmetry arises early in embryonic development; if it didn't, the outcome would be decidedly unlovely.

Early thinkers offered various explanations for how the body is shaped. An ancient Indian source declared the embryo to be fashioned from semen and blood, with the firm parts of the body coming from the father and the soft ones from the mother. In his *Generation of Animals*, Aristotle wrote that "the female provides the material, the male provides that which fashions the material into shape," in much the same way that a carpenter carves wood into a bedstead. Pliny (that Roman encyclopedist with the fatal desire to see the eruption of Mount Vesuvius) wrote this in the first century:

> Bears couple at the beginning of winter, and not in the usual manner of quadrupeds but both lying down and hugging each other; afterwards they retire apart into caves, in which they give birth on the thirtieth day to a litter of five cubs at most. These are a white and shapeless lump of flesh, little larger than mice, without eyes or hair and only the claws projecting. This lump the mother bears slowly lick into shape. (*Natural History*, Book VIII, p. liv)

At least Pliny gave credit to the bear. The medieval *Women's Secrets* presented the prevailing belief that heavenly bodies shaped human ones. In the third month of gestation, Mars divided the arms from the sides in the developing fetus, the neck from the arms, and formed the head. In the fourth month, the Sun created the heart. In the fifth, Venus perfected the ears, nose, mouth, and penis, and caused the separation of fingers and toes. Monsters of nature, caused by celestial influences, resulted either from too little heavenly matter or from too much; "in this way people are born with two heads or six fingers on one hand." (The book later notes that some monstrosities are caused by irregular positions during coitus: "I have heard tell that a man who was lying sideways on top of the woman during sexual intercourse caused the woman to produce a child with a curved spine and a lame foot.")

In the sixteenth century, certain natural philosophers claimed that they could detect in the head of a sperm a tiny person, a homunculus. This minute creature needed only to be transported into the woman's womb to grow. In their view, the first embryo of a species contained all future embryos — like those Russian dolls, ten wooden women in peasant dress, one within another, from the great bulbous Eve to a woman the size of a bean. Homunculi were nested inside homunculi *ad infinitum*, all the way back to the earliest embryo, a very small woman indeed. Development, then, was little more than a swelling up of the already complete being.

Paracelsus, the sagacious Swiss alchemist and philosopher of the great and small, held with the homunculus idea, and revealed his formula for creating one: "If the sperm, enclosed in a hermetically sealed glass, is buried in horse manure for about forty days and properly 'magnetized,' it begins to live and move. After such a time it bears the form and resemblance of a human being."

Over a century later, in 1677, Anton van Leeuwenhoek, who first saw sperm as parasites, seemed to accept this homunculus theory. Leeuwenhoek rarely described wonders that were not there, but one day, when he put a specimen of semen beneath the lens of a new and powerful microscope, he believed that he saw in the spermatozoon the outlines of a sort of embryo. A clamor of philosopher-scientists then claimed that they, too, had seen under the microscope minute forms of men in the semen of men, horses in the semen of horses, cocks in the semen of cocks. One went so far as to say that he had seen in a drop of donkey semen some very large ears.

The notion that a whole being — avian, equine, human — may be tucked into the head of a sperm, and, if lucky enough, may find an empty egg, climb in, lock the door behind it, and grow, is so appealing that it is easy to see why people were loath to abandon it. This was a generation that believed that mice arose from piles of old clothes, that geese sprang from barnacles, and that lower forms of life issued out of meat, mud, or slime. One could accept such magic in a self-contained universe — where the Earth was but a cup, the sky a cover, plants and animals frozen into their existing shapes; where the size of a creature had no lower limit. One could take life as it hatched.

To us, the analogies by which these early philosopher-scientists understood the birth of body shape are farfetched. Yet are they more so than the astonishing leap of self-organization that is the real genius of the embryo?

In the 1980s biologists zeroed in on some of the genes critical to this feat. One class, the Hox genes, shape the head-to-tail pattern of the body in the first few days of development, organizing it into front, middle, and hind region, determining the location of head, chest, and lower body, the general placement of limbs, digits, and organs.

The secret of Hox genes was revealed through a set of experiments with mutant fruit flies. The fruit fly, *Drosophila melanogaster,* originally came from a tropical region of Africa but has been cosmopolitan for some time, spoiling the bananas and peaches on kitchen counters and buoying the studies of biologists everywhere. The tradition of fly genetics goes back a hundred years to the discovery in Thomas Hunt Morgan's laboratory of a spontaneous mutant with white eye color. Morgan saw the fly, with its lightning-quick life cycle and prolific offspring, as a means for testing Mendel's laws of genetics. Morgan and his flies revealed the secret that genes are arranged on chromosomes in linear fashion, and that their order can be mapped by tracking the pattern in which genetic traits are inherited.

To understand how genes work in development, scientists expose their subject to X-rays or harsh chemicals that cause mutations and then note the effect on the descendants — how the progeny of the exposed creature grow, what sort of anomalies appear, such as normal body pieces growing in decidedly abnormal places or one organ changed into another. By looking at the irregularities in the offspring, they can deduce what the genes were supposed to do, and thereby determine their role in normal development. In the fruit fly, for instance, mutations in the *Antennapedia* gene cause legs to grow in place of antennae. Mutations in *Proboscipedia* make legs develop in place of probosces. These genes were called "homeotic" (from the Greek *homeo,* meaning "alike") because of their ability, when mutated, to transform one body part into the likeness of another. A mutation in one of these genes can cause radical change in a creature and spontaneous abortion.

When biologists studied these homeotic genes in detail they found, tucked within each, an identical fragment of DNA, sharply defined, as though enclosed in a box. They called the fragment a homeobox; the developmental genes that harbored a homeobox were called Hox genes. Now it's understood that Hox genes belong to a family of hundreds of different kinds of genes, all possessing a homeobox — a snippet of DNA that encodes a protein with one of those ancient keystone shapes, a homeodomain. This little "helix-turn-helix" twist can literally grab hold of the DNA inside a cell and control its fate.

As an embryo takes shape, each cell must know where to go and what to be, when to make specific chemicals — head-forming proteins, for instance — and when to shut down the production. When Hox genes switch on inside a developing cell, they help shape its identity. The protein they make screws itself into a groove in the double helix and switches on other genes that participate in the cell's development, setting off a cascade of biochemical activity that ultimately directs the body plan.

Since the discovery of Hox genes in fruit flies in 1984, more than a hundred such genes have surfaced in a broad variety of animals. Biologists "fish" for them, using the fruit fly genes as probes. To their astonishment, they have found Hox genes in sea urchins, worms, mice, birds, cows, humans, suggesting that a Hox gene or gene cluster existed in primitive form in an ancestor common to all living animals. (The genes have even surfaced in plants, although they're not thought to shape body plan the way they do in animals.)

The number of Hox genes differs from creature to creature. Humans have thirty-nine, most of them clustered in four sets on four different chromosomes; invertebrates such as fruit flies possess just one cluster of eight. But so close are these genes in form and function that when scientists performed the ultimate Frankensteinian experiment, inserting a human Hox gene into the embryo of a developing fruit fly, the human gene made a perfect little fly body.

And here's a surprise to vindicate those early philosopher-scientists and their beliefs about the origin of form. It turns out that Hox genes line up along the chromosome in roughly the same order as the parts of the body whose development they affect. Genes at one end of the line control the emerging head; genes in the middle, the abdominal segments; genes at the other end, the hind region. In a sense, then, the Hox genes represent a kind of biochemical homunculus tucked into the nucleus of early embryonic cells.

Hox genes don't act alone in determining body shape. Chemicals called morphogens, "makers of structure," working in the early stages of development, ooze slowly through the embryo, helping to establish compass points, tell head from tail, up from down, left from right. These morphogens are part of that complex system of communication among developing cells in the fetus. Some are the products of Hedgehog genes, so named because of the prickly look of fruit flies with mutated forms of these genes. (The names of developmental genes are nothing if not metaphoric, and create a rich taxonomic tree in my head: *late bloomer, hunchback, hairy, deadpan, daughterless, frizzled, wingless, lunatic fringe.*)

Hedgehog genes, like Hox genes, are common among different animals. The morphogens they make work along a gradient, diffusing as they ooze. Developing cells read the strength of these chemicals to determine where they are and what they should become — leg, wing, fin, brain. Biologists still have only a ghost of an idea how morphogens nudge a cell toward one fate or another, but they suspect the proteins stimulate other genes. One morphogen playfully named Sonic Hedgehog is secreted on the left side in a clump of embryonic cells. With the help of other whispering genes — *nodal, lefty, activin, snail* — it induces the left-right asymmetry of the vertebrate heart. One new theory suggests that this process is helped along by cilia, those whiplike hairs on the outer membrane of certain cells. The theory goes that the counterclockwise twirling of cilia creates a flow of morphogens that tells the embryo its left from its right.

Together, the proteins made by Hox, Hedgehog, and other genes create a significant chemical inequality that divides the embryo into front and back, bottom and top, left and right, limbs and head. It is this splendid chemical injustice, this early asymmetry, that we must thank for all the pattern we possess.

Imagine if one had to take conscious charge of this task. I'd run too many veins, drop whole skeins of nerves. I'd be asking whether this thimbleful of cells we had packed at the start were enough to make a whole limb. I'd worry that the yaw of that lung, the twist of this blood vessel, would get out of hand; shouldn't we shift that finger a touch to the left? No sculpture would be harder to shape than this.

Fortunately, an embryo generates the laws of its own becoming. With the help of a stable, conservative set of genes, the transformation from single cell to fully shaped being repeats itself precisely, generation after generation, in species after species, from worm to human. In fact, it is the rock-solid reliability of Hox genes that makes possible life's running sea of forms. If each new species had to reinvent the basic mechanism to control body pattern, there would be no time to evolve novel features, like wings, hooves, hammerheads, sicklelike jaws, serpentine tails, circles of whirling cilia. Nature improvises on a common theme.

The proliferation of primordial Hox genes half a billion years ago — or perhaps changes in the way they were used — may have sparked the explosion of complex anatomies in the Cambrian age. The genes must have evolved about the time body plans were diversifying, and their sequences have since held ground, like the body plans they help to create.

Some biologists suspect that a random mutation in the DNA of one of our ancient relatives half a billion years ago gave the creature a second set of Hox genes. The first set kept to its original role in shaping the body; the second set allowed the creature to evolve a fancier head, one packed with paired sensory organs and a complex brain — the forerunner, presumably, of our own massive lump of gray matter, with its specialized niche of tissue for animals, its need for nomenclature, its love of music.

For a long time it was believed that we vertebrates were born with our full complement of brain cells, and those that died were lost forever. (I've always thought it a delicious justice that birds are an exception to this rule. As they learn the songs of their species, new brain cells continue to bloom.) But lately has come promising news that our brains may indeed grow new cells throughout life, in the hippocampus and even in the neocortex, the part of the brain associated with learning and memory. Running boosts the growth of new nerve cells, at least in mice. And listen to this: rats engaged in the act of mothering sprout abundant new brain cells and do better than virgin females in tests of learning and memory, the result of hormones, perhaps, or the sheer metamorphic experience of motherhood. And, of course, while our neurons live, they go on making new connections or rearranging old circuits in response to experience. I like to imagine the molecular network created as I contemplate tiny, peninsular *Symbion* or consider the oozing diffusion of morphogens. Or when I look at my daughter's hands spread like starfish as she nurses, the twinned pairing of her limbs, bowed out in relaxation, the soft contours of her fat little symmetrical face.

7

NEW TRICKS

THE NOTION that Hox and Hedgehog genes explain the mystery of my baby daughter or, for that matter, a fruit fly is of course a fallacy. It's what I love about biology, about evolution, too; you can't narrow down the wonder.

I once saw a jigsaw puzzle created by a German biologist who studies the patterning of organisms from egg to being. The puzzle, made of 173 pieces of similar size and squiggly shape, formed a detail from a famous painting, a Brueghel. But in cutting the pieces along the intrinsic lines and colors of the painting, eliminating any connectors, Christiane Nüsslein-Volhard had made a puzzle so subtle and complex, I could hardly imagine where to begin.

It was Nüsslein-Volhard who showed how to systematically identify genes controlling development and, in so doing, unraveled the cluster of genes that establish the basic body pattern in the early embryo of the fruit fly; the work won her a Nobel Prize. The puzzle trick, however, she learned in childhood from her uncle, who delighted in devising games to stump his nieces and nephews. "You collect the pieces with similar patterns, *ja?*" she said to me, her fingers flying as she impatiently sorted the pieces: bricks, bark, snow. "Then you make another group and try to connect them." Under her expert hands, the subtle bulge of one piece tucked into the slight dimple of another, and hitherto unplaceable bits of serpentine gray and brown wood snugly filled gaps in a frozen lake, tree limbs, and heavy sky, gradually forming an image from Brueghel's *Winter Scene*.

It is the same with developmental genes. You have to find how the pieces fit together to shape the animal. In the 1980s, Nüsslein-Volhard set out to create a virtual encyclopedia of all the genes required to grow a fish. She and her team at the Max Planck Institutes in Tübingen, Germany, exposed male fish to the powerful chemical ethylnitrosourea, which randomly mutates the genes in all body cells, including sperm. Then they fertilized egg cells with the mutated sperm, and bred the fish for three generations. On examining the motley mutant offspring, they found thousands of aberrant fish, with flaws in nearly every facet of development: the growth and shaping of the brain, notochord, heart, blood, skin, eye, ear, jaw.

The team named the newly discovered developmental genes after the defects they caused in the fish: There was *lilliput,* which stunted growth, and *van gogh,* which caused malformed ears. *Santa* enlarged the heart; *moonshine* made fish glow in the dark under ultraviolet light; and *dopey, sleepy,* and *grumpy* caused defects in the nervous system. There were *spock, space cadet, riesling, rolling stones, bouillabaisse, slinky, sloth,* and two of Nüsslein-Volhard's favorites, *asterix* and *obelix* (named for the popular French duo in their striped pants), which gave the mutant fish a wild pattern of stripes.

These genes may be playfully named, but they have serious implications for our understanding of the way human babies develop and the nature of genetic flaws that maim or kill them. Some of Nüsslein-Volhard's zebra fish mutations — those affecting the heart, blood formation, and eyes, for instance — have counterparts among genetic diseases in humans. The mutant fish that failed to form blood cells in the usual fashion have the equivalent of some human blood disorders, including thalassemia and a type of congenital anemia.

What thrilled many scientists about the massive study was not the discovery of this or that individual gene, but, rather, the mapping of the genetic pathways, the sets of genes that click heels in the developing embryo to make borders, gradients, stripes, all the basic architecture of an organism. Genes nearly always operate in pathways to form the pattern of an organism. Regulatory genes at the top of the pathway make proteins that turn on other genes farther along, which make proteins that activate still other genes. "It's back to the jigsaw puzzle," Nüsslein-Volhard explained. More important than identifying any one particular piece is finding the connections between pieces and assembling them to configure the whole picture.

When news of the genes that set body pattern in fruit flies first whistled around the world, scientists went looking for equivalent genes in their favorite model organisms — mice, chicken, frogs. They were delighted to find parallel versions, not just of single genes but of intact strings of genes. There they were, those suites of genes that organized top and bottom, that shaped the asymmetrical organs of the early embryo (making the signaling molecules expressed asymmetrically to create the rightward looping of the developing heart or the gut coiled counterclockwise), that triggered the growth of wings, limbs, eyes.

Over eons, all variety of organisms have kept the same sturdy little gene pathways that operate to grant grids, segments, axes of up and down, right and left, near and far, mouth and anus, body and limb, organs and tissues. These sets of genes are the boats we all step into for the crossing from formlessness to form.

The gene pathways in different species may be alike, but they don't always do the same job. The one that sets up the basic body grid in the very early development of one animal may work later in the development of another — say, in the molding of limbs. Hedgehog and its partner, *wingless,* work in tandem to create body segments in the embryos of fruit flies. Hedgehog and a genetic cousin of fly *wingless* conspire to give the wing of a chick its root and tip, top and bottom, and to bestow human limbs with shoulder and fingers, palm and backhand, even to divvy up compartments in the human brain.

Nature may use the same developmental program in two disparate creatures but coopt it for a different job in each. The evolutionary lesson is that new form is born not necessarily from changes in the genes themselves, but from small, subtle shifts in the way they are deployed.

Here is the brilliant resurrection of an old idea.

There are minds, Emerson once remarked, "that deposit their dangerous unripe thoughts here and there to lie still for a time and be brooded in other minds, and the shell not to be broken until the next age."

In the midst of the French Revolution, the naturalist Étienne Geoffroy Saint-Hilaire got into an argument over the nature of animal body plans with Georges Cuvier, an eminent zoologist — an important debate in the history of science, described beautifully in a book by Toby Appel. In 1795, Geoffroy wrote:

> It seems that nature has enclosed herself within certain limits, and has formed all living beings on only one unique plan, essentially the same

in its principle, but which she has varied in a thousand ways in all its accessory parts. . . . The forms in each class of animals, however varied, all result in the end from organs common to all. Nature refuses to employ new ones.

A man of forceful manner and fighting temperament, Geoffroy believed that naturalists had falsely limited their investigations of the homologous parts of different animals: they had examined only cases where the shapes or purposes of the parts were obviously similar and had abandoned the search there. In so doing, they had failed to see that "nature works constantly with the same materials. She is ingenious to vary only the forms. As if, in fact, she were restricted to the same primitive ideas, one sees her tend always to cause the same elements to reappear, in the same number, in the same circumstances, and with the same connections."

Geoffroy published a provocative essay in 1822 in which he argued that the body plan of vertebrates was comparable to that of arthropods — insects, arachnids, and crustaceans. In considering a dissected lobster laid out on its back, he had been struck by its resemblance to a human lying on its stomach, the comparable organs in reverse order. "What was my surprise, and I add, my admiration, in perceiving . . . all the organic systems of this lobster in the order in which they are arranged in mammals?" To the shock of his colleagues, Geoffroy proposed a radical notion: the invertebrate body plan was but the vertebrate one turned upside down.

Cuvier, a specialist in comparative anatomy, believed that the great branches of the animal kingdom were absolutely distinct, marked by unique body plans that sprang into existence, perfect and complete, through the hand of the Creator. Cuvier was outraged by the suggestion that arthropods and mammals may share a body plan, writes Appel, that God may be "limited in His inventiveness" by a common design or collection of materials. As Cuvier saw it, the fingers of the human hand, the feathered wings of the bird, the claw of the mole, each was perfectly designed by the Creator for its particular and unique "office of existence."

For two months, says Appel, the men took turns defending their views at weekly meetings before a raucous audience at the Académie des Sciences in Paris. Not many cared to debate Cuvier. There is a portrait of him in the National Library of Medicine, looking cold, crisp, impatient, and capable of delivering a swift blow to soft thought. What could be more different, one from the other, than the body plan of a vertebrate and an invertebrate? Cuvier scoffed at the suggestion of unity, replying with a smug litany of differences between duck and squid.

In fact, few held with Geoffroy's wild hypothesis that arthropods and mammals have a common body plan — only flipped over — with its decidedly odd corollaries: that the legs of arthropods could be likened to the ribs of vertebrates, that the external skeleton matches the human spine.

But a "dangerous unripe thought" may lie latent, like a recessive gene, and spring to life under more auspicious circumstances. Lately, Geoffroy's theory has been unpacked and scrutinized again in light of startling new evidence. In the 1990s, developmental biologists fingered the genetic pathways in vertebrates that tell cells whether they're destined to become back or belly. In animals from frogs to humans, a pair of genes, *chordin* and *BMP*, work in tandem to tell cells to become the top of an organism or the bottom. In fruit flies and other invertebrates, a highly similar pair of genes, *sog* and *dpp*, work to the same end, only with opposite effects. The genes are equivalent but — as Geoffroy might have guessed — they act on opposite sides of the embryo. The gene that specifies fruit fly back is the same one that patterns frog belly. A human *chordin,* or "topside," gene can rescue a fly embryo lacking its *dpp,* or "belly" gene.

It's not that some ancient invertebrate experienced a Kafka-like moment in the backwater of deep time, waking from uneasy dreams to find itself transformed. But in an ancestor common to both vertebrates and invertebrates was a superb little genetic circuit that spelled out the belly-backside axis. Somewhere in the long pull of evolution, the circuit shifted slightly, inverting the body plan and setting life down a new path.

"Small changes in when and where the pathways are expressed have dramatic effects on body patterns," the biologist Sean Carroll told me. If a set of genes are expressed a little earlier or later in development, or if they're held in the *on* position a little longer, the adult animal will look vastly different, not just in patterning of body axes, but in shape and position of limbs.

New examples of this are cropping up everywhere. Consider *distal-less,* a handy, versatile regulatory gene for which Carroll has a special affection. As a child Carroll was fascinated by the bold patterns on the king snakes he kept as pets. Now an evolutionary developmental biologist at the University of Wisconsin, he is probing the genes that pattern appendages, from the hairy limbs of the fruit fly to the diaphanous wings of butterflies.

In studying what makes a creature sprout limbs, Carroll compared the development of legs in fruit flies and lobsters. Biologists had always assumed that the simple unbranched legs of flies had little to do with the complex branching limbs of such distant arthropod relatives as lobsters; that is, they arose from different ancestral structures and had no common genetics. But when Carroll looked at the genes involved in making cells bud from the main body axis to form limbs in lobster and fly, he found that *distal-less* sparks the event in both creatures; in the lobster, the gene is simply used twice to "bud" a branch from the existing leg.

William Blake once said, "He who prefers color to form is a coward." But now, Blake, it looks as if the two are closely linked, in their root genes at least. Carroll has discovered that butterflies use the *distal-less* gene not just to build their wings but to paint them with elaborate patterns. In this case the key is timing. The gene is turned on late in the development of the larvae and codes for the center of the butterfly's striking eyespots that flash from its wings, confusing predators or, perhaps, drawing mates. Moreover, the butterflies can use the same genes to change their wing patterns in just a few generations in response to a shift in predators or environmental stress. "Here's a remarkable example of how new patterns can evolve when old genes learn new tricks," said Carroll.

The appendages of all sorts of creatures, from worms and starfish to zebra fish and humans, start with a genetic "outgrowth" program involving *distal-less*. The gene is "old as dirt," Carroll explained. "The problem of developing something that projects from the body axis, be it claw or wing, was solved just once a long time ago, and that genetic mechanism is still at work. The architecture can vary tremendously, but the genes used to grow appendages are the same and have been preserved for a very long time. This doesn't mean that the limbs of an insect and a mammal are homologous, derived from a common ancestor, just that they're controlled by the same ancient genetic system, a 'sticky-outy' gene, if you will."

What differentiates crab from salamander, python from rhino, comes down to when and where the genes are expressed. Just such a redeployment of Hox genes during the development of a fin likely gave rise to a foot. In fact, Carroll believes this phenomenon may explain the great burst of new body plans during the Cambrian period. The eruption of strange, new forms, with their spectacular appendages, may have been the outcome of a "limbs race" among the Cambrian's many bizarre creatures, limbs for better sensing, moving, feeding, and grasping. The explosion was ignited, he reasons, not by a rash of new genes or genetic pathways but by new uses of flexible, successful circuits already in place in some ghostly creature at the base of the animal tree.

Imagine that animal, ancestor of sloth, butterfly, bobolink, dromaeosaur, and my own symmetrical body hatched in the plains of Nebraska. No fossils have turned up to show what this earliest ancestral animal looked like. "We're drawing a picture of something no one has ever seen," said Carroll.

Biologists once imagined this Ur-creature as a simple, smooth tube of cells, like a flat amorphous worm, with few distinguishing features. But now they wonder. If flies and mice share this set of highly flexible genes and gene pathways, then their earliest common ancestor must have possessed them, too — genes so successful and malleable they've survived to this day: Hox and Hedgehog for body shape, *distal-less* for appendages, *tinman* for some primitive precursor of a heart, Pax genes for eyes. Such a creature may have been far more sophisticated than a simple tube: bilaterally symmetrical, with a well-defined top and bottom, segments, a muscle that could pump fluid, some kind of outgrowth like an antenna, perhaps even an eyespot.

The sketch of their imaginings looks for all the world like Blake's *What Is Man?* drawing of the segmented, caterpillar-like creature — minus the infant face. It's not clear whether this ancestor actually had limb or heart or eye. But it must have been bursting with genetic potential, to creep along the warm mud, to move its own fluids, to sense sunlight it could not yet see.

8

SIGHTING LIFE

I HAVE BEEN TRYING to think of the shaping of an animal from egg to organism as a linear process, but it's no go. It is too subtle a puzzle, too messy and complex, with too many working parts tangled in. Take the work of three or four morphogens dispersing through an embryo, radiating their effects, and strings of ancient genes — Hox and Hedgehog, *tinman* and *distal-less* among them, gifts from some impossibly early ancestor — responding in mysterious ways for mysterious reasons, creating wave after wave of biochemical activity. The miracle is this: from the looping cascades of communication and control emerge the particular parts of a body in perfect form, nearly every time: the needle nose of the narwhal, the gossamer wing of a butterfly, the bony order of the snake spine, and — to my mind, most astonishing — the marvelous globe of the human eye somehow ready upon arrival out of a dark world instantly to receive light.

Not far from my house is a known and loved place, a path by a river where I go alertly, feasting my eyes on water and trees, where a chance happening — the mating of snakes, the tryst between orchid and bee, a good night view of faint stars — nearly always carves in my mind a splendid new molecular web.

Once while walking there I caught sight of a squirrel hung over the limb of a sycamore tree. I stood directly beneath the tree, my neck craned, waiting for the squirrel to move, but it hung there motionless, dead of unknown cause. From the high perch, a single eye, huge and white with cataract, gazed down at me. I wondered whether the creature had got by for a while with just this one eye aimed out at acorns and enemies, or whether its cyclopean existence had quickly done it in.

I walked on but could not rid my mind of that pale hailstone eye. It transformed itself, the way a mental image sometimes does when you try to flush it from your mind, into the keen blue eyes of a dog that had once lunged at me with bared teeth, then into the vacant eyes of my mother, dead from cancer, just before the nurse brushed her lids shut, then into the wide-open terrified eyes of an injured bird, from somewhere in my childhood, which I thought necessary to put out of its misery. I remember holding the weapon, a stone, and the bird's eyes following my movements, the summer sun glinting off its corneas. I closed my own eyes but still couldn't bring myself to carry out the task.

Eyes — the sweet wet wink of an otter, the pink demoniac beads of a rat, the slot-eyed stare of an octopus — eyes are what we hold most in common with other animals, what draws sympathy and fear. Eyes have been called exact witnesses, lights of the body, best artist, best composer, and window to the soul — a phrase that sent me straight to the mirror when I first heard it, searching for the organ of my spiritual being perched invisibly behind my iris.

Nearly every kind of animal has some kind of eye, from the eyespot of a microscopic tardigrade — a simple affair of five or so light-sensitive receptor cells, which can tell the creature about the distribution of light and darkness but cannot form images — to the sophisticated "camera-style" eyes of humans and octopuses, which can see images of brilliant clarity. There's the fiercely focusing eye of a hawk, a kind of telephoto lens that gives the raptor vision twice as sharp as our own and enables it to pick out a rodent scuttling in the underbrush a thousand feet below. The huge, sensitive, camera-type eye of the Australian net-casting spider *Dinopis* allows the creature to snare its prey in the Bible black of a forest at night. The eyes of *Anableps*, a Caribbean fish that swims along half in the water and half out, are bifocal; each eye has two pupils, one suited for looking upward to watch for predatory seabirds, the other adapted for gazing down into the water to hunt for prey. Then there are the twenty-four complex image-forming eyes of the cubomedusoid jellyfish *Tripedalia*, a creature with no central nervous system to process or interpret complex visual images — a biological conundrum if ever there was one.

Here's an image: the sun setting over the river, the water alive with striders that dimple its calm surface. A mimosa tree extends one long limb over the water; hovering above its pink spray of blooms is a shimmer of humming-birds, their iridescent green breasts winking in the low rays of the sun.

That a pair of two small orbs, accounting for less than 1 percent of the weight of my head and asking only 1 percent of my cardiac effort, can take in all this — dazzling light, subtle color, lightning-flash movement — impresses me.

The ancient Greeks believed that the eye had the probing power of touch. Plato wrote in *Timaeus* that when the gods assembled the body, they put in "the vessel of the head . . . a face in which they inserted organs to minister in all things to the providence of the soul. . . . And of the organs they first contrived the eyes to give light." Plato held that particles streamed from the eyes like probing rays, touching an object and thus making it visible. To Euclid and Ptolemy, it made sense that light traveled from the eye, that all sensation came from direct contact between the perceived object and the organ of sense. But Aristotle attacked the idea: "It is unreasonable to suppose that seeing occurs by something issuing from the eye; that the ray of vision reaches as far as the stars." Objects emanated rays, which impinged on the eyes with the help of air. As Cicero described it, "The air itself sees together with us." Even Galen, who knew the anatomy and physiology of the eye, knew the cornea, iris, vitreous and aqueous humors, and retina, and identified the crystalline lens as the chief organ of sight in the eye — "a fact clearly proven," he wrote, "by what physicians call cataracts" — even Galen thought intervening air an instrument of vision.

It was Abu Ali al-Hasan ibn-al-Haytham, born in 965 in what is now Iraq, who explained the mystery of vision by inviting observers to note how the sun scorched the eye. His book on optics became a standard guide to the theory of light and vision in medieval Europe. But not until the seventeenth century was it fully accepted that what we see is not the bird itself but the light reflected by it.

Whatever enters my eye — squirrel, strider, hovering bird — enters as light only, as a beam of photons, tiny particles of packaged energy. What those particles meet is a constellation of coordinated parts. They pass first through the transparent cornea, curved and clear, bulging from the white like the glass on a watch face, which reduces the speed of light and bends it toward the center of the eye. Then through the port of the iris, which governs the amount of light admitted by continuously adjusting the aperture.

Then into the lens, a finely crafted crystalline structure with a ruff of muscles that contract and expand a hundred thousand times a day, allowing the eye to focus first on something an inch or two from the nose, next on Venus. Then those photons penetrate the clear jelly of the vitreous humor, turning once like the roll of a seal. Finally they strike the retina (from the Latin for "net"), a screen the size of a postage stamp with more than a hundred million photoreceptor cells — cones for color in bright light, rods for vision in dim light. The waiting photoreceptor cells, loaded with light-sensitive pigments, transform the light energy into quivers of electrochemical energy that travel to the brain, which edits them into recognizable images.

The steps in the sequence take a long time to describe but only fractions of a second to execute. On the one side, there is the dense, bony darkness of our internal being; on the other, brilliant image. Eyes are the instantaneous interface.

When Charles Darwin first presented his theory of evolution, the human eye was used as a favorite example to point out the weakness of the theory. How could this complex system of perfectly synchronized and integrated parts have come about little by little? How could cornea, lens, retina — not to mention the three sets of muscles that move the eyes back and forth, up and down, several times a second, and the fine circuit of nerves that links all of these components and controls their actions — how could these multiple parts, arranged in perfect geometry, have been spontaneously assembled over time by the blind force of natural selection? Wasn't it far more likely that the fabulous eye popped into existence all at once, a creative act of God?

The riddle of how such an exquisite organ could have arisen by chance gave Darwin himself a cold shudder. In a chapter of *Origin of Species* entitled "Difficulties of the Theory," he wrote:

> To suppose that the eye, with all its inimitable contrivances for adjusting the focus to different distances, for admitting different amounts of light, and for the correction of spherical and chromatic aberration, could have been formed by natural selection, seems, I freely confess, absurd in the highest possible degree.

But in the end he found he could accommodate even this miracle within his theory:

> If it could be demonstrated that any complex organ existed, which could not possibly have been formed by numerous, successive, slight modifications, my theory would absolutely break down. But I can find no such case.

Step by step, through the accumulation of small changes over time, a simple, imperfect eye is transformed into a complex one.

In the early 1990s two Swedish scientists, Dan Nilsson and Susanne Pelger, set out to determine just how much time such a transition might take — in, say, a worm or mollusk or other smallish aquatic animal of the kind that may have been alive in the Cambrian age. The earliest fossils of animals with eyes date back to this period, 550 million years ago.

As a starting point, the scientists assumed the existence of a primitive light-sensing organ composed of three layers: a flat, clear surface, light-sensitive cells beneath, and, below them, dark pigment. They used a computer model that allowed the top, transparent layer to deform itself randomly 1 percent at each step. For a change to be accepted by the model, it had to improve the amount of spatial information the eye could detect. The scientists used conservative rules about genetic change and heritability, overestimating the time required for a change to take hold in a natural population.

Their 1994 paper, "A Pessimistic Estimate of the Time Required for an Eye to Evolve," is astonishing. The model eye deformed itself in time-lapse imagery on the computer screen with lightning speed, the flat light-sensitive patch dipping first into a shallow cup, then into a deeper cup, and finally a sphere with a lens, a full-fledged eye capable of forming its own image. The number of steps required to progress from light-sensitive patch to complete camera-style eye was only 1,829. Given a generation time of a year or so (a conservative estimate for the small marine animals at issue), it would take only a few hundred thousand years to evolve a decent complex eye. Since the Cambrian, enough time has passed for eyes to evolve fifteen hundred times — and that's just in one lineage.

The trick apparently gives nature no bother. Biologists who have traced the evolution of different visual systems in the animal kingdom believe that eyes evolved at least forty times in different lineages, probably as many as sixty times. I like to imagine this great revolution in vision, eyes opening everywhere: the five globular eyes of the Burgess Shale's *Opabinia;* the bulging multifaceted compound eyes of the dragonfly; the colossal pie-like eyes of ichthyosaurs, twenty-two centimeters across, the better to see in the dark of the deep sea; the bulbous eyes of alligators popping up through duckweed scum; the quick black eyes of a lizard; the keyhole eyes of the green whip snake; the peduncled eyes of crabs and lobster; the long, tubular, highly sensitive telescopic eyes of the deep-sea fish *Scopelarchius,* which point straight up through black water to gather the dim silhouette of prey against residual sunlight.

The oldest known eye, that of a fossil trilobite, is a compound eye as big as a dragonfly's. At least half of all animals, including most insects, have compound eyes. Unlike the human eye, with its single lens and single retina, a compound eye is an intricate arrangement of many little eyes, or ommatidia, each with its own retina and lens. A firefly's eye has several hundred ommatidia; a honeybee's eye, several thousand.

Compound eyes confused early researchers, who thought that insects must see images repeated many times over. But it turns out that the insect sees a single complete scene, as we do, with each ommatidium contributing its portion of the image, like a pixel. But because image quality depends on lens diameter, and the individual lenses of a compound eye are very small, insects likely see a more blurred image. Humans resolve images about a hundred times better. Insects, however, outdo us at detecting movement.

At last count, scientists had turned up at least nine different optical systems in the natural world, among them, simple pinhole eyes, eyes built like mirrors or like reflecting telescopes, two kinds of camera-lens eyes, curved-reflector "satellite dish" eyes, and several kinds of compound eyes. With the possible exception of the zoom lens and one or two other optical systems, it seems that nearly every method of producing an image has been tried in the animal kingdom — some of them highly unusual.

When the British biologist Michael Land looked through a dissecting microscope into the bright blue beadlike eyes of a scallop, he saw an inverted image of his room, including a distorted picture of himself looking through the microscope. A master of eye research, Land realized that the image the scallop sees is formed not by a lens but by a spherical silvery mirror lining the back of the eye, the optical analogue of a Newtonian telescope.

Then there's the unusual eye of *Copilia quadrata,* a rare aquatic creature found deep in the Bay of Naples, the smallest known creature capable of forming an image. In the nineteenth century, Siegmund Exner, an Austrian naturalist and expert in optics, described the miniature marine arthropod as an exquisite transparent creature about the size of a pinhead, with a pair of lens-like structures deep within the body that "showed the most lively movements." When Richard Gregory, British psychologist and vision researcher, stumbled on Exner's description more than a century later, he was so intrigued that he went to the Bay of Naples to search for the creature:

> Each day the laboratory's boat-men brought us a large glass jar, densely packed with all manner of plankton hauled from the deep water just beyond the Bay. Each day we searched, drop by drop, by eye and with low- and high-power microscopes. . . . After about two weeks we almost gave up. Then suddenly — there she was!

Very square, with a pair of perfectly contrived anterior lenses that looked like car headlamps and, deep within her transparent body, another pair of lenses — pear-shaped — moving away and toward each other in a sawtooth scan. Here was an instance of a scanning eye, the kind of optical system used in a television camera.

In Italo Calvino's book *Cosmicomics,* a mollusk takes credit for creating the first eye by creating the first image worth seeing, a striking spiral shell bright with colors and stripes — which, by force of its beauty, succeeds in "digging" the eye and its tunnel to the brain.

The secret of how evolution first cupped out the eye's small windows is still sleeping in nature. So wildly different are the eyes of the animal kingdom that scientists have always assumed they must have arisen from vastly different beginnings. To be sure, all eyes have in common certain structural components. Ubiquitous in vertebrate eyes are crystallins, the proteins that make up the lens. So perfectly arranged are they to bend photons toward light, so finely tuned to their task, that it was believed they are highly specialized, evolved solely for their role in seeing. Then came the news that one kind of crystallin was in fact the very protein the body used as a common "housekeeping" enzyme for the plebean metabolic tasks of the cell. The vertebrate eye lens, it seemed, had committed a kind of genetic piracy, enslaving a common enzyme, refining it a little, and repackaging it to make the perfect lens protein.

The eye lenses of fish, scallop, strider, hummingbird — indeed, of all animals — are guilty of the same trick. The mix of crystallins may differ from animal to animal, depending on the particular requirements of different environments — in sea or desert, in dim light or bright — but the proteins themselves were recruited from a set of common enzymes.

Or consider rhodopsins. These proteins act as the visual pigments of rod cells that detect light in all multicellular animals; they even guide eyeless green algae and bacteria toward or away from light. One evolutionary geneticist has fingered the slight molecular shifts in rhodopsins that affect a creature's ability to see in the diverse conditions of different habitats. By comparing the sequence of rhodopsins in fish that live near the ocean surface with those in coelacanths, which thrive in the blackness of the ocean deeps, the scientist found the amino acid substitutions that changed the spectral sensitivity of the proteins — a neat tracing of adaptive change right down to its molecular roots.

But despite these common components, the eye of a mammal and the eye of a fly are about as unlike as any two things can be, from the shape of the organ and how it comes about during embryonic development, right down to the molecular mechanisms underlying the workings of its cells. And while the single-lens eyes of the octopus and human are similar in basic structure and function, they develop in such different ways in the two embryos that they, too, have been pegged as the descendants of different ancestral eyes. In fact, eyes have been considered a premier example of the phenomenon known as convergent evolution: different creatures hitting on a similar solution to a common problem but from remote starting points.

Down by my river path not long ago, I got a lesson in convergence. Out of the corner of my eye, I saw a winged thing hovering above a flowering shrub. It looked for all the world like a hummingbird, though it lacked the distinctive markings of our native ruby-throated species, and my pulse quickened at the thought that I was witnessing a rare stray or accidental. A closer look, however, revealed that the creature had no beak but a proboscis, a long flexible tongue that uncoiled to probe the flower's interior. It was, in fact, an insect, a common hawk moth, the mature form of the tomato hornworm that plagues our crops. I'm not alone in having been so fooled. In *The Naturalist on the River Amazons,* Henry Walker Bates wrote:

> Several times I shot by mistake a humming-bird hawk-moth instead of a bird. . . . It was only after many days' experience that I learnt to distinguish one from the other when on the wing. This resemblance has attracted the notice of the natives, all of whom . . . firmly believe that one is transmutable into the other.

That creatures as unrelated as moths and birds have settled on the trick of hovering in midair while drawing sweet sustenance from the nectaries of deep-throated flowers — the hummingbird, with its long thin bill, the moth, with its proboscis — is a beautiful case of convergent evolution.

The notion that all the various eyes of the natural world were classic examples of convergence seemed to be upended in 1995 by a strange and ingenious set of experiments. A group of Swiss biologists, led by Walter Gehring, were studying how a gene called *eyeless* helps build the compound eyes of fruit flies. When the team found a way to switch on the gene in the different tissues of a developing fly embryo, they were startled by the results: the flies hatched with perfectly formed eyes in out-of-the-way places — on wings, knees, even on the tips of antennae. Turning on one gene in a tissue that would normally make a wing or an antenna had made that tissue into an eye. Some flies sprouted as many as fourteen eyes. The scientists knew that it takes hundreds of genes to make an eye, but the *eyeless* gene appeared to possess the masterly ability to trigger the event all by itself. As Gehring said, it was like finding a single gene that would turn a human heart into a liver.

Then came another surprise. Scientists had earlier discovered that mice and other mammals harbor an eye-making gene called *Pax-6.* When the Swiss group put a mouse *Pax-6* into the different tissues of the fruit fly, the mouse gene performed just as the fly gene had. Eyes popped up in decidedly un-eyelike places. The reverse worked, too. The fly version of the gene made eyes develop in strange places in mice. *Eyeless* and *Pax-6,* it turns out, are members of a family of genes with a homeobox, that little bit of DNA that makes a protein capable of binding to, and switching on, a whole battery of other genes, in this case, eye-forming genes.

Humans also have a version of the mouse *Pax-6* gene. Mutations of the gene in humans may result in the birth of a child with eyes missing an iris, that rare congenital condition called aniridia. No iris means no control of the light entering the eye, and people with aniridia often suffer from cataracts, glaucoma, and poor vision. It was a mutated *Pax-6* gene in that nineteenth-century Canadian woman that started her long familial string of seventy-seven descendants with aniridia.

So similar are the human and mouse versions of the *Pax-6/eyeless* gene, 92 percent the same, that *eyeless* might produce secondary eyes in us, too. The gene may be universal in animals. Even the nematode *C. elegans* has a version, though the nematode has no eyes. The gene helps pattern the worm's head region and a sense organ in its tail. *Pax-6/eyeless*, then, may be an old gene, born in some Precambrian creature, where it shaped the fate of cells in the nervous system.

News of these shared master eye genes doesn't contradict the view that sophisticated, image-forming eyes evolved independently many times over. But it does suggest that insects and mammals somewhere in the distant past shared an ancestor with eyes built by the ancestor of the *Pax-6/eyeless* gene. Those eyes perhaps were primitive, simple eyespots — or patches of light-sensitive cells like those that respond to light in the genitals of butterflies — present in an early common ancestor, an ancient relative of Italo Calvino's lucky mollusk, an Ur-creature hovering in the dark, ready to catch seeping light.

Richard Gregory tells the story of a patient, blind since infancy, who, when he was fifty-two, had his sight restored through a corneal graft. Much to the surprise of his physicians, the patient could tell the time by a wall clock immediately after the operation. Even more surprising, he was able to read capital letters flashed before him, although he could not recognize lowercase letters. As it happened, before the operation, he had used a large watch without a glass cover to tell the time with his hands. At a school for the blind, he had learned capital letters by touch but was not taught lowercase letters. Somehow, the patient instantly understood visually the things he had learned through touch. His image of the tactile had leaped to the visual.

Scientists suspect that a leap like this may have given birth to sight. At some point in deep time, through a chance mutation or two, a nerve cell sensitive to touch, or a pigment cell in an early version of skin, caught a break from its dark past and fired off at vibrations of light. In a sense, then, the Greeks were right: sight may have been a species of touch — at least in its earliest beginnings. Later, sight itself might spark skins of beauty. Zoologists have found that certain animals present in the Burgess Shale flickered and flashed with a splendid iridescence that arose, they believe, with the emergence of eyes.

9

THE CATERPILLAR FACULTY

IN SCULPTING body and mind, in building eye and nerve, skin and bone, muscle and blood, cells bulge to nearly twice their volume, split, differentiate, and swell in numbers, a feat of growth and exponential increase that deserves the greatest respect.

But something else is going on as well.

I once watched a paleontologist work with a thin needle to liberate the fossilized remains of an ancient bird from its matrix of volcanic tuff. When the scientist first found the rock, he could see the wispy outlines of a vague form. Over the course of months, he slowly picked away the bits of ash and rock until a set of fine bones and feathers emerged — beautifully articulated skull, beak, wishbone, tibia, wings.

The body's limbs and organs undergo a similar kind of chiseling. The hand begins as a paddle-like paw. During the sixth week of embryonic life, the webbing between thumb and pointer, ring finger and pinky, is whittled away. Sex organs, too, are shaped in this way. In males, the cells of the Müllerian duct are carved out (these would otherwise form the female's uterus and oviducts). In females, it's the cells of the Wolffian duct that are eliminated (and would otherwise form the vas deferens and other male parts). So, too, in a developing brain, neurons bloom in great profusion; then 90 percent of them are winnowed out.

That development in certain organisms involves some death and dissolution of body parts has been understood for centuries from observations on metamorphosis. I once chanced on a collection of brilliantly colored paintings of South American insects, some of them nightmarish — a giant cockroach scaling a pineapple, ants swarming over a branch while nearby an enormous hairy tarantula devours a hummingbird — but most

of them showing the metamorphosis of moths and butterflies from egg to
winged insect, the life stages presented as if they were happening all at once
and so exquisitely beautiful they took my breath away.

The paintings were the work of Maria Sybilla Merian, a naturalist and
artist who left her home in Amsterdam in 1699 at the age of fifty-two and
sailed to South America to paint the life cycles of insects in the jungles of
Suriname. The journey would have been remarkable in any century, but
was especially so in the late 1600s, when savants were still poking for an-
swers to nature's riddles in the ruins of the classic world, quoting the au-
thority of Pliny (who had confidently proposed that butterflies were born
from dew), and drawing their knowledge of animals from medieval besti-
aries. Birds were categorized according to notions of nobility, from most
noble (eagles and hawks), to wise (owls), to big (ostriches), on down the
line to boobies, stupidest of the lot. Caterpillars, if they were considered at
all, were classified apart from winged insects and lumped together with
worms and serpents. These lowly beasts crept into paintings mostly as symbols: fly as sin, butterfly as resurrected soul, worm as a creature whose pri-
mary purpose was to confront mortal man in his coffin. Women of the time,
if they traveled at all, traveled in the company of family or large groups.
Here was a woman moved to take a long and dangerous journey by the
sight of a single butterfly specimen from a distant land.

Not many understood what errand Merian was on.

From Paramaribo the artist sailed up the Suriname River, stopping
along its banks to collect caterpillars, then watching them closely for signs
of metamorphosis and painting them at the pivotal moment. Two years
later, sick of heat and fever, she returned home, loaded with brandied butterflies and chrysalises, lizard eggs, and hundreds of paintings of iguanas
and geckos, fighting snakes and frog-eating scorpions, and dozens of moths
and butterflies in various life stages on their native cassava, guava, batata,
and pawpaw plants.

Almost forty years earlier, when she was thirteen, Merian had begun a
journal in which she described the unbounded joy she first felt at seeing
caterpillars spin cocoons and metamorphose into moths. Later she wrote
of one elusive moth: "When I caught sight of it, I was enveloped in such
great joy and so gratified in my wishes that I can hardly describe it. Then
for several years in a row, I got hold of its caterpillars and maintained them

. . . on the leaves of sweet cherries, apples, pears, and plums. . . . Strange to note," she wrote, "when they have no food [they] devour each other."

Obsessed with the transformation of insects, she devoted herself to caterpillars, raising them in her yard, collecting them in the pleasure gardens and weed patches of Frankfurt, where her father's business was, and, later, on the heaths and moors of Friesland, in the moat at Altdorf, the pleasure gardens of Nuremberg, and the jungles of Suriname.

That an adolescent girl should focus her light on metamorphosing Lepidoptera is not so surprising. I raised no caterpillars in my garden, but I do remember being fascinated when I was twelve or thirteen by the Merlin-like evolution of moths and butterflies from lumpish grub to airy winged thing. It was that moment in life when the boyish image I had of myself blew apart, and I had no idea what to replace it with. I loved the idea of a creature emerging from the prison of its larval state radically transformed, sloughing off an old identity to assume with confidence an entirely new one. I sought out the myths of metamorphosis, tales of the weaver Arachne, who hanged herself and was changed by Athena into a spider; Actaeon, the feckless hunter who was turned into a deer by Artemis and torn apart by his own hounds; Proteus, who tried on every identity from stone to serpent. I devoured stories of shape-shifting witch doctors, medieval necromancers, shamans, those subject to magical deaths, then dismemberment, then resurrection in a new body created for them by the spirits.

The first real insect metamorphosis I observed took place in a small acrylic box under a large striped tent in the parking lot of a shopping center near my home in Virginia. Above the tent flew a bright yellow banner announcing the annual Butterfly Fair. Inside, the air twitched and skittered with meadow browns, copper hairstreaks, cabbage whites, brimstones, and skippers, so airy that they seemed less solid beings than winged ideas.

I stood before the plastic box willing the little pouch to disgorge its treasure. I had seen numerous caterpillars in the wild, of course, and, once, what I took to be a swollen woolly bear caterpillar ready for a great transformation. But I'd never witnessed the moment of passage. That afternoon in the parking lot, while I stood shifting my feet on the hot macadam, reading the educational poster next to the boxed chrysalis, fanning myself in the warmth of the tent, it happened just as I wished: the sac split, and a shiny, tight little scroll shouldered out of its pupal prison, moist and vulnerable. It hung limp until it was dry, then struggled to expand itself, and cracked into a delicate spread of dewy medallioned wings.

Francesco Redi, the personal physician to the Grand Duke of Tuscany, proposed the notion in 1668 that insects were born not from dew or decaying flesh — as Aristotle and Pliny had believed — but from eggs deposited on plants or meat. Here was as good a case as any of art inspiring science. Though Redi had built his reputation on his observations of vipers, the physician was also versed in literature. It was a passage from the *Iliad* that sparked Redi's doubts about spontaneous generation. In the nineteenth book, Achilles asks his mother, a goddess, to protect the corpse of his friend: "I much fear that flies will settle upon [him] and breed worms about his wounds, so that his body, now he is dead, will be disfigured and the flesh will rot." To probe the point, Redi placed pieces of meat in eight flasks, sealed four, and left the others open to the air. Maggots quickly infested the meat in the open flasks; none appeared in the ones that were sealed. A year after Redi's discovery, Marcello Malpighi, the Italian biologist who fathomed the cellular nature of plants, strode in to explain the transformational nature of insects from egg to winged thing.

If we are to believe her journal, Maria Sybilla Merian knew the details of metamorphosis perhaps more intimately than these savants. Her paintings and journals address the step-by-step transfiguration not just of Lepidoptera but also of amphibians. In 1686 she had precociously noted the developmental odyssey of frogs from black grains that "fed on the white slime that surrounded them," to tiny creatures that grew little tails so that they could swim, then eyes, then, eight days later, two little feet "from the skin at the back and after a further eight days two little feet at the front . . . [like] small crocodiles. Thereafter the tail rotted away and they became proper frogs and jumped onto the land."

Just what was going on in this rotting away wasn't discovered until 1842, not long after the discovery of animal cells. That year a German biologist, Carl Vogt, peered through a microscope at a developing midwife toad, *Alytes obstetricans*. Vogt saw that the individual cells of the toad's notochord were being "resorbed," as he put it, swallowed up or sucked in. But he didn't think his observations important, and they slipped into oblivion.

Twenty years later, scientists observing metamorphosis in flies, ants, and beetles noted in the changing muscles and glands the wholesale death of cells. Later studies of developing bones, muscles, and other tissues in mammals suggested that the shaping of ear, eye, nose, tongue, intestinal tract, and trachea involves some removal of excess cells. By the turn of the century it was clear: to have shapely toads, butterflies, babies, one must have cell death.

The idea that death would chaperone growth was not easy to accept. But at least the role of cell death in the miracle of development seemed of minor importance. Then, in the early 1970s, a team of Scottish pathologists stepped in to darken the picture, suggesting that in the shaping of body and mind, cell death is massive.

The team had examined different kinds of tissue and discovered that cells have two radically different ways of expiring. Those that die accidentally, by injury or poisoning, swell up and pop like balloons, spilling their insides over neighboring cells and inflaming nearby tissues — a visually obvious process the team called *necrosis*. But those that die naturally during development do so by quiet, efficient, nearly invisible means. They shrivel up before they leak their insides, then dissolve away from their neighbors, which quickly dismember the dead and eat them up in a neat and orderly manner, leaving no corpses. This way of dying (which the team called *apoptosis* from the Greek for "a falling") may be inconspicuous, but it goes on all the time in every kind of tissue, in nearly every kind of animal.

Cell death is the night-side of growth. Cells die *en masse* at incredible speed, not just to sculpt our body structures before we are born, but throughout our life, about ten billion each day, in blood, brain, intestines, skin, uterine wall — to adjust cell numbers; eliminate the injured, the malignant, the malfunctioning; and dispose of cells deemed obsolete. We don't diminish in the face of such massive, incessant loss for the simple reason that new cells are born in trade for those that die. By some mysterious means, the body precisely balances cell birth and cell death.

To make cells and then quickly discard them may seem an extravagant waste. But overkill is common in life. Nature aims above the mark to hit the mark. I think of the 30 percent of proteins summarily destroyed within minutes of their birth, the nine million eggs in a single spawning of one Atlantic codfish, a man's superabundant production and loss of sperm. Fecundity and waste go hand in hand. Nature pours it on, burgeons, prolifer-ates, runs up and over, endlessly leafing and flowering; to every creature, even the smallest, adding a surfeit, a generosity, a drop too much, of sheer delightful waste.

Over the years, various theories have been put forth to explain why cells die. Perhaps they simply exhaust their "life energy." Perhaps as an organ's shape changes, the cells are pushed aside and robbed of their nutrition. In the late 1980s, the biologist Bob Horvitz made a bold suggestion: cells die as a natural result of the process of growth because they have a built-in pro-gram to take their own lives. Just as cells carry within them the seeds of their propagation, they also harbor the means of their end, a little program to dismantle their lives, to commit suicide.

A cell's decision to be or not to be often depends on the chemical sig-nals it gets from its neighbors. Consider that squidlike nerve cell from Di-ane Hoffman-Kim's film. When a nerve cell is born, it sends out axons, like fine tendrils, to forge connections with other nerve cells, muscles, or sen-sory organs. It follows a scent trail of protein messages sent by the cells it is supposed to contact. If the axon follows its correct route, it gets the signal to survive; if it strays, it dies. Once linked to its target tissue, it receives a steady flow of signals reassuring it of its importance. If the tendril senses none of these so-called savior signals, the cell body of the neuron automati-cally activates its suicide machinery.

This may explain why the body makes billions and billions of cells only to ask them to die before they've had a chance to act. Producing an overabundance of cells, which must compete for limited amounts of sur-vival signals secreted by neighboring cells, is a way of selecting for the "best" cells.

To live or to die, then, is often a social question — which is why scien-tists have such difficulty keeping individual cells alive in a petri dish. With-out neighboring cells around to whisper "live," only a witches' brew of chemicals will keep an isolated cell from taking its own life.

The means by which a cell commits suicide was largely opaque until Bob Horvitz took some lessons from the nematode worm. As the worm matures from egg to adult, exactly 131 of its cells choose to die, each at its carefully controlled, appointed time. To find out which genes may be in-volved in the grim task of self-sacrifice, Horvitz and his colleagues at the

Massachusetts Institute of Technology first treated nematodes with a chemical that causes random mutations in some genes, and then looked for mutant worms with a defective cell-death program. Two genes played a role in all 131 suicides. If either of the two failed at its job, the worm held on to all of its cells.

Both genes have human counterparts that participate in the multiple suicides going on in our own bodies. One gene makes a protein that chops up the proteins of a cell's insides — its nucleus, endoplasmic reticulum, and cytoskeleton — until, in the words of one biologist, the cell suffers "death by a thousand cuts." The second gene makes a protein that boosts the first gene's ability to slice and dice.

In *C. elegans*, Horvitz found another gene that suppresses cell death. If that gene is knocked out, most cells commit suicide, and the worm dies early in development. It, too, has a close human counterpart, so close that when scientists put the human gene into a developing worm, it halts cell death. When this human gene is mutated within the human body, it can cause certain forms of cancer.

It's not that the molecular dance of death is identical in worm and man. In our tribe, cell death is more complicated, with cells integrating many more signals before deciding to live or die and activating a dozen or more genes in a long, complex genetic pathway to perform the deed. But many of the genes are the same, and the program works in the same way, through scores of carefully choreographed steps, not just in worms and humans, but in fruit flies, flowers, slime molds, even some bacteria, such as *E. coli* — which suggests that the genes for cell death emerged early, perhaps in a single cell more than a billion years ago.

Now there's food for thought: On the surface of it, a single cell programmed for suicide makes no biological sense. Death for one means death for all. One theory suggests that the program has deeply heroic roots: it emerged in some bacteria as a mechanism for self-sacrifice to protect neighboring members of the same strain from viral infection. A bacterium infected by a virus might commit suicide before the virus reproduced, thereby saving its genetically identical siblings.

"The thought of suicide is a great consolation," wrote Nietzsche. "With the help of it one has got through many a bad night." But Albert Camus called suicide the "one truly serious philosophical problem. All the rest —

whether or not the world has three dimensions, whether the mind has nine or twelve categories — comes afterwards."

The idea of cells taking their own lives for the sake of other cells may seem unnatural, especially if one's inclined to value the individual or believe life forms are driven by purely selfish interests. But examples abound of animals giving up their lives, most often for the sake of their kin — worker bees that attack intruders to protect their hives (their sting is barbed, so their pulling away from a victim is an act of pure, unqualified suicide); plovers that risk their lives to distract predators from their nests; naked mole rats infected with parasites that wander out of their burrows and starve themselves to avoid infecting their kin. One of the more peculiar instances of self-sacrifice occurs among the dipteran fly known as a midge. The mother harbors her offspring not in the protective envelope of a uterus but within her own tissues, which the young devour from the inside out, eventually consuming all. Within forty-eight hours, they, too, are offering their insides to their children.

Cell suicide makes good biological sense in our bodies, to scrap the webbing between fingers, to excise tails and other vestigial structures, to winnow down the seven million egg cells made by a female fetus to the four hundred thousand a woman possesses by puberty. Or to thwart *horror autotoxicus,* an autoimmune disease, by culling cells from the developing immune system that carry receptors capable of reacting against the body's own cells. All for the good of the whole growing organism.

But poising cells on the edge of suicide, making them rely for life on the whispered words of neighbors, seems exceedingly risky. If cell death is sparked by accident, crucial cells may die before their time. In a developing fetus, alcohol can trigger apoptosis in millions of neurons in the forebrain just as they're forming connections, and this can cause severe problems later in life, such as learning and memory disorders, depression, and psychosis. Scientists learned lately that if an infant mouse is deprived of its mother's attention for a single day, its brain cells commit suicide at twice the rate of the cells of its steadily mothered counterparts.

To protect against accidental cell death, pieces of a cell's apoptotic machinery are sequestered in different places — in the membrane of the cell and in its mitochondria. (It makes an interesting footnote that these bacterial descendants, once thought simply to generate energy for my cells, also help dictate whether my cells live or die.)

Tightly governed though it may be, apoptosis is a hair-trigger affair. I once asked my friend Laura Attardi why nature places cells on the brink of suicide, given the risks.

Cell death keeps cell growth from getting out of hand, she told me. "A good brake is as important as a good motor." Laura pronounces her name the Italian way, with a wide-open first syllable and a lovely rolling *r*. Her father came from Padova and named his daughter after Petrarch's love. Laura began her career studying gene expression. Then, a few years ago, her stepfather — her guardian and mentor — died of pancreatic cancer at the age of fifty-three. Now she focuses her light on a gene that promotes cell death, a gene named, by coincidence, p53.

François Jacob once said that the "dream" of every cell is to become two cells. Growth is a lust for cells, an insatiable hunger. Think of the dividing and pumping and growing of all flesh, the beefing-up and burgeoning, cells dividing every second in the body to replace lost legions, not just in the developing animal but throughout life. "The problem," Laura explained, "is that every dividing cell is potentially a tumor cell." Programmed cell death is in part a hedge against nature's ominous tendency for extravagance. When a single cell with damaged DNA becomes deaf to the whispered urging of its neighbors to check out, it may violate the code and proliferate wildly, spiraling into massive, excessive growth. Growth can easily become overgrowth. Laura's stepfather died of such an aggressively growing cancer. So, at the age of fifty-one, did my mother.

The p53 gene has a range of duties. In the brains of young mammals, it is active in pruning out the vast oversupply of neurons as the connections between brain cells are refined. In organisms from flies to humans, it works to keep cell growth in the body from getting out of hand by binding to the DNA in a cell and instructing it in proper behavior. Called the "guardian of the genome," p53 prevents a cell with damaged DNA from dividing until the cell has had time to repair the damage. If the wound is too severe, the gene sentences the cell to suicide, purging the body of the seed cells of cancer.

The ability to die on demand is the bargain that individual cells make in order to live and grow harmoniously in a single organism. It's a kind of pact, like the one in some societies where old people, having reached a certain age, distribute their belongings, close their doors for the last time, and walk into the wild, making room for the young.

·   ·   ·

I have been trying to imagine life without this cellular death wish. There would likely be rampant cancer and paddle-like hands and unformed sex organs and heaven knows what other deformities of excess. When scientists knock out the cell-death genes in nematodes, they find that the worms hanging on to those extra 131 cells don't function as well as their leaner counterparts. But they do live a full life and die of old age. However, flies deprived of their cell-suicide genes, *reaper, hid,* and *grim,* die early in development. So do mice, their central nervous systems clogged with a superabundance of nerve cells. Since people are genetically closer to mice and flies than to worms, we can suppose a similar end.

I know now that cell death is necessary for doing away with excess nerve cells and thereby solidifying in my brain the billions of connections that give rise to pleasure in pondering the metamorphosis of a silkworm or the making of an eye. I know, too, that the body makes more brain cells than it needs so that it may select the best, the ones that successfully find their targets. We filter out superfluous cells just as we filter out all but a small percentage of the sights, sounds, and other sensations around us so that we might pinch meaning from the few. Out of cell death comes birth, growth, language, memory, vision, mind enough to wonder. Wasn't the ancient Egyptian symbol for life a serpent swallowing its own tail? Wherever you enter the circle, it is dying; follow it, it is growing.

10

SEXING LIFE

OUT AMONG THE WET GRASSES by the river one day, I saw two snakes engaged in what looked like a ritual dance. They coiled and uncoiled in a lean, undulant double helix, at times so tightly wound they seemed one beast with two heads. I was flat on my belly for a snake's-eye view, mesmerized by the slim twisted lengths. Here is a body blueprint stripped and minimal, free of ears, whiskers, limbs, the better to slip and pour along the ground. A look at the deep evolution of the snake body plan reveals this new trick: a radical change in the expression of Hox genes, resulting in limb loss and the development of a long supple spine with which to heap and wind and writhe.

I think of snakes as cold and cautious, but these two, twining naked in the morning dew, were pure sensual elegance. Nearly identical in size and coloring, they were wonderfully homologous ropes of muscle and nerve, like rippling, looping mirror images of each other. I assumed they were coupling male and female, but they could as easily have been two males in combat.

Male and female snakes typically resemble each other. Sexing them is a skilled profession. Michael Grace, a neurobiologist at the University of Virginia and an amateur herpetologist, once gave me brief instruction in the task. We were in a small sunny room at the back of his brick house in central Virginia, home to thirty serpents, from racers and whip snakes to pit vipers and boa constrictors, one of which, housed in the next room, had a severe cold and from time to time sneezed like an old man.

Size and color rarely offer clues to a snake's sex, Michael told me. As he spoke, a rainbow python moved up his arm, around his neck, down the other arm, then straight out into space. Every so often, Michael shifted his hands, smoothing down the length of the snake, picking up the thick slack between thumb and forefinger. The rainbow python is not poisonous; it's a constrictor. This one was five feet long with a sumptuous pattern of gold-and-chestnut spots, and, when its scales caught the light, a stunning prismatic sheen like the glaze of a katydid's wing.

The shape of the tail can offer clues to gender, Michael explained. Males often have longer tails, a dimorphism that arose early in evolution: the tail of the male rainbow python is long and thick to the end; the female's tapers down in lovely sinuous diminishment. But I was distracted. When the snake moved in my direction, I shrank back a little, stung by a fear ancient and powerful, rooted in the psyche of my species, a mystic horror and superstitious loathing hardwired through generations of experience with venomous species and often played out in stories, epithets, metaphors: forked tongue, snake pit, nest of vipers, low-down as a snake's belly, cold as a coiling snake, crooked as a snake's path, dread viper, subtlest beast of all the field. "We have it from many authorities," wrote Pliny, "that a snake may be born from the spinal marrow of a human being." What more chilling thought. People in diverse cultures dream about snakes more than any other kind of animal; in many cultures they are central symbols for the continuity of creation and for genitals.

I had to squelch my fear. The more reliable method of sexing a snake involves inserting a slim metal probe into the snake's cloaca. If the probe slides in less than a quarter of an inch, the length of a few scales, you probably have a female. Males have a deeper pocket; it will swallow up to two inches of probe, depending on the species. Michael showed me how to lubricate the probe with water and slide it in slowly, toward the tip of the tail. It's a delicate procedure. If I pushed too far, I risked injuring the snake.

To sex an alligator, you can use your hand, sliding two fingers inside the slit to feel around for a hard slippery length. The method is not failsafe, however; the female has a sizable clitoris, which can make for confusion. The lower shell, or plastron, of a male turtle sometimes has a slight invagination that fits over the back of a female; the female's plastron is flat or convex.

A crow can sometimes be sexed by size, the male being slightly larger than the female. Though the birds known as blue tits are sexually dichromatic, the color difference lies in the ultraviolet part of the spectrum, which birds can detect but we mammals cannot. (For the bird, it's an appealing trait; a female blue tit prefers a male with more ultraviolet light reflecting from its crest.) In many species of birds, no size or plumage differences distinguish the sexes, as far as we know. Only lately have ornithologists discovered a universal marker that betrays the gender of nearly all birds, a gene they can extract from a single molted feather. Marine biologists have found a similar way of sexing humpback whales from the DNA extracted from skin sloughed by a swimming whale.

With many animals, however, from bees to bulls, peacocks to people, gender is betrayed by looks — flamboyant plumage, long tail, horns, antlers — or by voice — the whine of a katydid, the ascending trill of a prairie warbler, the visceral croak of a frog. In humans, a difference in pitch of an octave or more conveys the sex of the speaker, the result of a larger larynx in men and a longer more massive pair of muscular flaps, or vocal cords. Of course it's generally not necessary to sex a human by vocal pitch. There's facial hair, chest hair, breast size, shoulder-to-hip ratio, relative height, relative stature, and, yes, genitals.

The English language has whipped up an astonishing wealth of words for the organs that distinguish males and females. I recently found an impressive little list that science devised to describe the complex male genitals of crickets and their allies: phalli, epiprocts, paraprocts, cerci, gonotremes, and titillators. But nothing holds a candle to the verbal virtuosity displayed in the naming of human sex characteristics. For female breasts, there's bubbies, boobs, orbs of snow, milk-shops, dumplings, dairy arrangements, globes, charms, and heavers. For penis, *The Slang of Venery* documents a sort of etymological family tree of more than six hundred synonyms. In *Mrs. Grundy,* an entertaining study of words plain, prude, and rude, Peter Fryer lists those terms found in this and other dictionaries:

arbor vitae
best leg of three
bit of stiff
prick, cock, pile-driver
Jock, Jack, Willie, Robin, Roger, Peter, Dick
finger, dagger, dart, spear
pike of pleasure

flap-doodle

silent flute

sensitive plant

hermit

sex

shaft of delight

poker, gooser, tickler, shove-devil

nightstick, joystick, sugar stick

lollipop

sweetmeat

banana, potato finger, live sausage

mole, mouse, worm, bird, goat, snake

pee-wee, pendulum, P-maker

rump-splitter, bush-whacker

trouser snake

thumb of love

waterworks

wombbrush

tail, tool, unruly member

Nimrod, Nebuchadnezzar, Abraham, Old Rowley

Dr. Johnson

(The last name found its place on the list, says the *Dictionary of Slang*, because there was no one whom Samuel Johnson was not prepared to stand up to.)

Now, in the full scope of English vocabulary, the hundreds of words devoted to the male sex organ are a drop in the bucket. But when you compare the lexicon of slang devoted to other bodily parts of *Homo sapiens* — say, torso, stomach, limbs, and head — genital anatomy seems to soak up more space in the brain than its relative physical proportion should call for. This, says the scholarly Peter Fryer, reflects the verbal inventiveness, sexual vigor, and pride of Englishmen over several centuries. (Or perhaps, as Pliny explained it, a poignant preoccupation: "All the other animals . . . experience satiety in coupling, but with man, this is almost entirely absent.") It is also evidence that we consider the presence or absence of the male sex organ to be the essence of sexual difference, the fundamental, defining feature of life's two basic models.

So distinct are the secondary sex characteristics of men and women that it's a shock to realize that we all start out with pretty much the same equipment. Penis and clitoris derive from the same organ, as do scrotum and labia majora. Early in development, we all possess the same indifferent gonads, the same simple ducts. The human embryo stays in a mode of indecision for nearly two months before its gonads commit to one sex or the other, and its organs and ducts diverge. Seminal to all, too, is the genetic blueprint for both male and female bodies. With the exception of a few genes on the sex chromosomes, every functioning gene in a man's body is also found in a woman's, and vice versa. Both men and women possess a gene that controls penis length; only in male bodies does it express this effect.

I like the notion that the human genome harbors two "interleaved texts," as the biologist Robert Pollack described it: "one to be read only in male cells and the other only in female cells. . . . It speaks to the subtlety of our genomes that all of us have DNA sequences our bodies will never read — sequences that can be read only by cells in a person of the other sex." At least at the start, we are both male and female within the small boundaries of nearly all of our cells — with one notable exception: our gametes.

Here's the secret of the true distinction between the sexes. Males make small sex cells; females, big ones. This is the one common feature that can be used throughout the natural world to label males as males and females as females.

Even if the difference is visible only under a microscope, a two-sex system is the law of most life. Given this — given how deeply embedded in nature is the making of gametes and the sexual way of reproducing — it comes as a surprise that the genetic means for making sexual difference varies wildly from creature to creature, even among those we consider to be close relations.

When I am out of joint from a poor run of thoughts, I like to page through the notebooks of Leonardo da Vinci, to think of him feverish with musings on why fossil shellfish are found on high mountaintops, on the nature of moonlight and sight ("why a picture seen with one eye will not demonstrate such relief as the object seen with two eyes"). His notebooks are marked by a kind of orderliness, on the one hand ("begin the anatomy at the head and finish at the sole of the foot"), and a glorious chaos on the other, a hodgepodge of anatomical drawings, physiological diagrams, maxims, morals, polemics, fables, jests, architectural and mechanical sketches, and notes in his curious mirror writing:

> describe the tongue of the woodpecker and the jaw of the crocodile
>
> which is the part in man, which as he grows fatter, never gains flesh?
>
> get coal
>
> describe why water moves, and why its motion ceases . . . and how water rises in the air by means of the heat of the sun, and then falls again in rain; why water springs forth from the tops of mountains
>
> supreme fools [are the] necromancer and enchanter
>
> the cause of tickling
>
> the cause of lust
>
> I wish to work miracles

In the third volume of his notebooks, *Quaderni d'Anatomia,* is Leonardo's work on embryology: beautiful sketches of a dissected pregnant uterus and its membranes, anatomical drawings of the embryo side by side with pictures of weights and pulleys, observations on embryonic growth ("the length of the umbilical cord always equals the length of the foetal body in man though not in animals"), and this note:

> Eggs which have a round form produce males, those which have a long form produce females.

The shapely egg theory of sex determination was launched by the ancients and perpetuated by a host of followers: "When you would feast upon eggs, make choice of the long ones," advised the poet Horace; "they are whiter and sweeter and more nourishing than the round, for [these latter] being hard . . . contain the yolk of the male." That males come from more spherical eggs and females from more oval ones accorded with the beliefs that the sphere is the most perfect of all figures in solid geometry, and the male the more perfect of the two sexes.

Democritus held that female embryos originated from the left testis, males from the right. To procure the more perfect gender, Anaxagoras recommended that males lie on the right side during sex. Left and right figured in Hippocrates' belief about sex determination, too, but his concerned placement in the womb: right for male, left for female. For news of a child's gender, said the medieval *Secrets of Women,* the expectant mother should note the hints in her anatomy: for a male child, an abdomen rounded and protruding on the right and a bigger right breast; for a female child, an abdomen of an oblong shape, pain on the left side, a left breast blackened, and the milk it produces, bluish and watery.

Aristotle held that sex was determined not by side but by the heat of the male partner during intercourse — hot semen made males; cold, females — and advised men to copulate in the summer if they wished to have heirs. Females were but "mutilated males," whose development was cut short by a cold womb. Étienne Geoffroy Saint-Hilaire, champion of unity in body plans, also held with the importance of environment, but he believed that both sexes were present in the embryo at conception; the determination of sex depended on the size, shape, and motion of reproductive organs, and the quantity and quality of nourishing fluids. Followers of Leeuwenhoek and other "animalculists," who regarded spermatozoa as the source of all new life, believed sperm the key to gender. One could see beneath the microscope "spermatic animalcules" of both sexes, distinguishable, like snakes, by slight differences near their tails.

It's true, of course, that we humans owe our sex to the fortune of a single sperm. In breeding his fruit flies in the 1930s, Thomas Hunt Morgan discovered that sex is determined at the moment of fertilization through the meeting and pairing of sex chromosomes. The human egg carries an X chromosome; the sperm, either an X or Y. If the sperm delivers an X, the egg will produce a female; if a Y, it will produce a male.

The Y chromosome looks like an upside-down version of the letter for which it's named and is the smallest of all mammalian chromosomes, harboring only two or three dozen genes. The X — which owes its name to the obscurity of its nature at first discovery — is by comparison huge, roughly six times larger than the Y, and rich with thousands of genes, among them many involved in brain development. (That is why mutations in the X frequently cause mental retardation.) The size discrepancy is what allows scientists to "sort" male and female embryos, thus enabling people to choose the sex of their infant.

There was probably a day deep in time when the X and Y were equals.

Scientists suspect that their ancestors came from an identical pair of standard chromosomes about 350 million years ago, just after mammals split from a turtlelike reptilian ancestor. Then, early in the evolution of mammals, the Y began to degenerate; now it's a ghost of its former self. Some scientists believe that it may be doomed to extinction. But, for now, it still carries a gene that delivers the punch of maleness — a fact confirmed by exceptions.

The word "hermaphrodite" comes from the names for the Greek gods Hermes and Aphrodite. According to Greek mythology, a celestial union of the two brought forth a child who merged with a nymph and became a body endowed with the attributes of both sexes. The cleavage or bisection of hermaphrodites often figures prominently in stories about the origins of human males and females. The theme is nearly as widespread as the mythological power of the snake.

Several animals are hermaphrodites. Some contrive to be both sexes at once, such as slugs, sponges, jellyfish, and worms, which line up belly to belly with the heads pointing in opposite directions and discharge sperm into each other's bodies. Other creatures change sex, starting out as one and switching to the other. Oysters do this regularly at intervals throughout their lives, with the sex organs ripening alternately. A fish called the blueheaded wrasse plays the game another way. In a school of wrasse, all of the fish are females except the largest, which is male. If he dies, within minutes the largest female begins to behave as a male. After a couple of weeks, she will go through a physical transformation and become a male.

Though true hermaphroditism — possession of both testes and ovaries — is rare among humans, having a mixture of male and female characteristics is not. About one in every twenty-five people blur or bridge the great divide. In the early 1980s scientists began studying the sex chromosomes of people with sex reversals — men who had the female genetic XX pattern but were nevertheless male, with the full range of secondary sex characteristics, and women who had an XY pattern but were still female — a relatively rare disorder, striking about one in twenty thousand. The scientists discovered that the Y-less men did in fact harbor a piece of the Y chromosome, the very piece that the XY females were missing. This squib of DNA held a key to sexual difference. But was there a "maleness" gene? And if so, where was it?

That squib of DNA turned out to be mighty long, more than 300,000 base pairs. One promising gene within it turned up in a snake called the banded krait. The gene also appeared on the sex chromosomes of mice, birds, and fruit flies, but not humans. Then, in 1990 scientists zeroed in on another candidate. When this gene was injected into newly fertilized mouse eggs, genetically female mice emerged from the womb with a decidedly male appearance. The scientists named the gene *Sry,* for sex-determining region of the Y.

Just how the *Sry* gene performs its symphony of services has yet to be fully fathomed. Like a Hox gene, it may be the master switch inside the cell, binding to DNA, bending it, and radically altering its properties during the eighth week of pregnancy. In so doing, it turns on a slew of other genes that signal the body to transform its gonads into testes and thereby initiate the biochemical cascade that ends up in maleness. The testes churn out the potent hormones that spur the growth of male organs and destroy tissues that would otherwise become female organs. Ducts that might have become fallopian tubes and ovaries degenerate; those that form the vas deferens grow.

Sry doesn't act alone in making males. For one thing, a gene on the X chromosome gives the male body its ability to react to androgens, the so-called male sex hormones. And the female path is no "default" mode. Body types, both male and female, are the result of a multitude of changes brought about by hundreds of thousands of genes, located on different chromosomes. Some of the genes are under the spell of hormones circulating throughout the body, which ensures that all cells receive the same signal and develop in the same sexual mode. But even here nature abhors a monopoly. In humans, both sexes have both "male" and "female" hormones coursing through their blood. Androgens are now known to be made in the ovaries and to trigger the growth of pubic hair in girls. Testosterone sculpts secondary sex characteristics in boys and in girls. Both estrogen and estrogen receptors are in the fluid inside the testes and may be essential for male fertility. Male mice lacking a certain estrogen receptor are infertile.

David Crews, a zoologist at the University of Texas, has urged scientists to focus less on the differences between the sexes and more on their similarities. In his view, the two sexes are much alike, with inseparable roots. The female was the ancestral sex, the first self-replicating organism; it gave rise to the male, a variant, and the two still share many characteristics.

When it comes to sex, we all have a doubleness within, and more fluidity than some of us would like to admit.

While the *Sry* gene has been around for hundreds of millions of years and is still busy in birds, reptiles, and bony fishes, it's only in mammals that it plays a role in determining sex. Scientists suspect that it may have evolved from a gene on the ancestral X or Y chromosome that had some other purpose in life, perhaps shaping the liver or brain, but in mammals was co-opted for the purpose of making the male sex.

Unlike other genetic pathways in development — for body segments, for eye or limb or heart — the mechanisms controlling the differences between male and female differ radically across the animal kingdom.

Although the heat of semen at the time of copulation doesn't result in maleness in mammals, as Aristotle proposed, temperature does determine sex in some reptiles. The sex of a baby crocodile or turtle depends on the heat at which the egg develops. Temperature shapes the abundance of hormone receptors in the growing embryo. Unlike reptiles, birds use genes to determine sex, but it's the males who have a pair of identical sex chromosomes, ZZ, and the females who have one large and one small, Z and W. A Z chromosome is nothing like a mammal's X, which suggests that they evolved from different chromosomes altogether. The mechanism of sex determination in birds remains a mystery.

From the thicket of sex-determination differences in the animal kingdom scientists lately snatched a small similarity. A gene on Chromosome 9 that plays a role in the shaping of human testes also has a part in shaping the different sexes in birds and in *Alligator mississippiensis*. This gene may represent an ancient conserved piece of the vertebrate pathway of sex determination. But in general, so vast is the difference between a human's way of determining sex and a reptile's or a bird's, that it almost equals the gulf in their physical appearance. Even within the close-knit mammal family there are startling departures in sex-making strategies. In marsupials, such as opossums, one X codes for a scrotum; two Xs result in a pouch and mammary gland. Female lemmings often have the XY pattern of sex chromosomes. Certain voles possess no Y chromosomes at all.

The notion that sex comes about in diverse ways even within a single genus suggests that sex-determination mechanisms are fickle, apt to evolve rapidly. In fact, scientists have lately created mutant strains of the nematode worm *C. elegans* with an entirely different sex-determination system than that of the parents. If a laboratory geneticist can bring about this kind of change in a few generations, imagine the possibilities of natural change acting on large numbers of individuals over many generations in all varieties of species.

By diverse means we arrive at the same end. How marvelous that the mechanisms for creating sex differences are so flexible and fluid. How strange that they so often create two sexes.

Why only two? Why not three or four or twenty? That way one's choice of partners might not be limited to just half of the population.

One explanation for life's popular two-sex state has been proposed by the evolutionary biologists Laurence Hurst and William D. Hamilton. In mammals, at least, it may result from the constraints of our mitochondria, those small strangers to our flesh. When two gametes meet, they bring to the union not just the DNA in their nuclei, but also the little bacterialike loops of DNA in their resident mitochondria. The nuclear DNA from gametes mixes together nicely, becoming two pairs of chromosomes. But mitochondria don't like to mix genes or share cytoplasmic space, and because they possess some rather nasty enzymes, they can avoid doing so. If a struggle for mitochondrial dominance were to erupt, a zygote might suffer serious damage.

The theory suggests that the great bilateral split, now so ubiquitous in the natural world, may have evolved, in part, to avoid massive mitochondrial wars. One sex — the male — unilaterally disarms. When a sperm fertilizes an egg, it obligingly surrenders most of its mitochondria at the door, donating for perpetual inheritance only its nuclear DNA and shedding the cytoplasmic rest, like a snake sloughing its skin.

When it comes to sex, we inhabit a mystery. Erasmus Darwin, grandfather of Charles and a physiologist in his own right, called sexual reproduction "the masterpiece of nature." Hurst wrote that it is "one of the most risky endeavours performed by eukaryotic organisms." If the goal is to pass our genes to the next generation, then wouldn't cloning do it a great deal better than sex? The animal kingdom does possess its asexual species — about two thousand of the planet's two million or so named species. Among the most notable are the bdelloid rotifers, tiny translucent sacklike animals that live in fresh water. The bdelloids have abstained from sex for close to forty million years. They reproduce parthenogenetically, the females laying eggs that develop without fertilization into more parthenogenetic females, and so on; no males have ever been seen.

But birds, bees, people — indeed, most known organisms — depend on the more risky, demanding route of sex. Why is puzzling, as it costs dearly in time and energy to make sex cells; to find and woo desirable partners through extravagant and boasting feathers, songs, calls, contests; and, finally, to join compatibly.

Take wooing. Anyone who has ever watched a pair of squirrels skittering in fits and starts about a tree trunk or male hummingbirds hovering, plummeting, arcing up in acrobatic flight to seduce a mate, wonders at the energy expended. Guppies swoop in on predators just to dazzle potential mates. The fastest muscle of any vertebrate belongs to the male toadfish, *Opsanus tau,* which uses it not for fleeing predators or capturing prey, but for courting. It contracts and releases the muscle surrounding its gas-filled swim bladder two hundred times a second, just to whistle at passing females.

Or the sex act itself. Giant squids engage in boisterous acts of sex in the dark of the deep sea, the male using its muscular three-foot penis to inject sperm packages under pressure directly into the appendages of females. A male red-back spider gives up the game altogether, positioning himself just above the female's jaws during copulation, and afterward submitting to sexual cannibalism.

As for compatible joining of genes, Hurst wrote, "The very bringing together into one territory (the zygote) of unrelated genomes, should result in tragedy, like the union of the Montagues and Capulets."

Why does so much of life bother with sex? To evolve quickly, says one theory, to keep pace with predators and parasites by making it difficult for them to penetrate bodily defenses. ("It takes all the running you can do," the Red Queen told Alice, "to keep in the same place.") To dump harmful mutations, says another. When two sex cells fuse, one may be able to compensate for the genetic weaknesses of the other.

My favorite theory belongs to David Crews: "Biological systems require complementarity. Hormones need receptors, parasites need hosts, one sex needs another to interact with in complementary sexual behavior patterns."

So, too, said Emerson: "There is somewhat that resembles the ebb and flow of the sea, the days and nights, man and woman, in a single needle of the pine, in a kernel of corn, in each individual of every animal tribe."

The first organisms may have floated alone and isolated in an ancient sea, but life has never been alone since. The energy to split, to draw boundaries, do battle, is matched by the tendency to pair, to join, link and fasten, to depend and support, to dance naturally toward a companionable kind of life.

Part III

RELATION

A logical or natural association between two or more things; kinship, connection; also, the act of telling. From the Indo-European *tel-*, to lift, support.

11

THE MOTH BENEATH THE SKIN

IN HIS 1877 ESSAY "Biographical Sketch of an Infant," Charles Darwin noted that his one-month-old son "perceived his mother's bosom when three or four inches from it," and responded by fixing his eye on her breast and puckering his lips. Darwin doubted the response had anything to do with vision or touch; he suggested that the baby was able to sense the warmth and odor of his mother's breast.

It's true. When she's nursing, my baby daughter is most akin to a small blind mammal driven in my direction purely by scent. She may well have been exposed to this odor before birth. The olfactory system is already in place in a fetus at twelve weeks, long before other sensory systems emerge. The womb holds a rich mixture of chemicals — simple sugars, amino acids, small peptides, hormones — which are transported across the placenta to circulate in the fetal bloodstream. From here, they may pass into the amniotic fluid, where the fetus can taste or smell them. In fact, our first sensory perception may be the smell of our closest relative at the moment, the odor of the fluid in our mother's womb.

Sight may be ancient and pervasive, and touch, too, the sense by which the locust knows the wind. But smell, the ability to sense chemicals, is most often the way mortal beings detect the presence of someone or something outside themselves, something other. Smell is as old as life itself and present everywhere, even in the "lowliest" organisms — often with an acuity that exceeds our own. A whiff of truffles buried in soil and oak duff three feet deep will make a boar foam at the mouth. By flicking its forked tongue, a rattlesnake can sample the strength, left and right, of subtle chemical signs deposited by would-be mates and prey. On this playing field, the advantage nearly always goes to the visiting team. In fact, if one were assembling a file of "Big Differences," a compendium of traits distinguishing humans from other members of the family of life, Sense of Smell might head the list.

One spring morning my dog Lucy shoots out the door and down the path to the garden, nose to the wind. Suddenly she wheels around, twitching, electric, full of excitement as she probes a spot in the dewy grass. Part bird dog, Lucy is small, near the ground, privy to its secrets. She has caught an aromatic vapor from unseen origins, and she lingers for a moment, savoring, then snuffles forward to seek other faint effluvia. When she finds a scent to her liking, she rolls herself over it from tip of nose to end of tail, turning first on one side, then on the other, to rub herself with this secret emanation. Then she gets up, shakes herself, and trots off to another cloud.

I suspect that Lucy classifies her finds in a complicated taxon subdivided by type and time: new or old, fresh or putrid, fragrant or stinking, happy, delicious, brave. I wish I could get inside her brain and know a world so arranged. I have tried to learn from her, to pry sensual secrets from the soil. I stoop to snuff out her news, but to my nostrils nearly every spot is neutral, devoid of any appreciable olfactory character. I inhale deeply but can smell nothing more than wet earth tinged with a fungal mustiness and — now that she's leaning into my leg to encourage me — Lucy's own sweet doggy blend of warm grass, sunshine, and crushed leaves, along with the new element, a faint fecal note impregnating her damp coat. But this is a crude gleaning on the surface of a deep olfactory mystery, a rich, subtle chemical text that I can barely discern but that Lucy absorbs in full. I come back from her school humbled, feeling the mediocrity of my equipment and the poverty of my skills.

Lucy's nose has about two hundred million olfactory receptors, the neurons active in smell. That's twenty times the number in mine, and she's probably a thousand to a million times more sensitive to odors. A trained bloodhound can distinguish between the smell of two pennies minted in the same year, between two scent trails left hours earlier by men wearing shoes — which is why dogs are adept at snuffling the earth for trails, sniffing out firearms, drugs, and disaster victims, and why the United States Department of Agriculture has a Beagle Brigade to patrol airports for gypsy moths, snakes, and contraband mangoes and clams.

It's the same with other creatures. A lobster can sift through a chemical dictionary among the currents, reading with exquisite precision the scents of life and death by way of two wands jutting from its head. It can do this even in turbulent, murky bottom water, where a scent plume is quickly washed out, just as the skywriting of an airplane is dispersed by the wind into unintelligible little puffs of smoke. Birds use odors to sniff out food, select good nesting material, navigate across vast stretches of land and sea. Circling vultures have been used to spot leaky gas pipelines; the birds are drawn by a chemical in the gas with a scent like carrion. Spawning eels and salmon find their way back from the open sea, up rivers and tributaries, to streams where they were hatched, guided by a mix of watery scents imprinted in their memories.

Even these talents pale in comparison with the skills of a moth. Early one evening in 1875, the French entomologist Jean-Henri Fabre caged a single female great peacock moth in his laboratory and left the window open to the night. When Fabre later opened the door, he found the room a wild flutter of rare wings, like "a wizard's cave with its whirl of Bats." Over the course of eight days, a hundred and fifty moths came from afar, from within a radius of perhaps a mile and a half, "apprised I know not how," he wrote. Endeavoring to show that smell was *not* the key, he filled the room with suffocating fumes of naphthalene, the aromatic component of coal tar. Still the males came. Fabre could not bring himself to believe that the female was able to advertise her presence over so great a distance by odor alone, since "one might as well expect to tint a lake with a drop of carmine."

It was a good analogy. A male of the silkworm moth *Bombyx mori* can sniff out air tinted with a quadrillionth of a gram of this odor — a scent females secrete to announce their sexual readiness — and can use it to home in on a mate though she be several miles downwind.

Dog, lobster, snake, moth, flea: most animals live and love by the nose and are superbly equipped to glean what's crucial in the world by reading its chemical messages. For them, air, water, and soil are odor clouds packed with a babel of urgent words circulating in wild confusion, changing in composition, character, and intensity, which they wander into, sample, and find meaning in, catching the whiff of edible morsel or poison, the odor of threat, passport, invitation, call to arms, trail home, precise boundaries of real estate. From this buzzing, blooming chemical confusion, dog and moth receive unequivocal directions; sniff out the fragrance of friend, enemy, mate; regulate their own internal state by reading the internal state of another. Who's in my family? Who's in my group? Who's dominant? Who's ready to mate?

That smell is an essential element in the lives of nearly all animals, single-celled creatures included, even Ernst Haeckel knew. In *The Riddle of the Universe,* the nineteenth-century German biologist wrote that the main cause of attraction between two sex cells "is a chemical sensitive action of the protoplasm, allied to smell and taste, which we call 'erotic chemico-tropism'; it may also be correctly (both in the chemical and in the romantic sense) termed 'cellular affinity' or 'sexual cell-love.'"

We seem to be failures at this game, biologically unfit for picking up on the chemical drift, or, perhaps, above it all, having evolved out of a need for scent to guide hunger and love. Are we hopelessly outside the circle of this private language? Wadded with stupidity and chemically deaf?

Smell has long occupied a bottom spot in the sensual hierarchy of philosophers. The eye and the ear are noble, said Plato, because they sense geometry and music, worlds of perfection. The nose, on the other hand, is a lowly, sordid organ, tuned to the foul emanations of body and breath. Always inclined to typology, Plato divided smells into two categories, the pleasant and the unpleasant. The latter "roughens and does violence to the whole cavity lying between the crown of the head and the navel; the pleasant soothes this region and restores it with contentment to its natural state." Three hundred years later, the Roman philosophical poet Lucretius made the distinction even more physical: "The things which are able to affect the senses pleasantly consist of smooth and round elements," while bitter and harsh particles are hooked and "wont to tear open passages into our senses."

According to medieval physicians, unpleasant odors caused disease and pleasant ones chased it away. When the plague struck fourteenth-century Europe, the disease was thought to be caused by foul odors and was fought with an arsenal composed largely of sweet ones. Bonfires of pine and fir were lit in the streets — one for every eight houses — and were kept burning day and night to drive away the pestilent stench. Physicians made their rounds in long coats of black leather, their heads encased in masks with long beaks full of fragrant herbs and dried petals to ward off the disease. Around the house of the dead was sprinkled scented plague water, the origin of the term "eau de cologne."

Immanuel Kant called smell at once the most necessary of senses and the most unproductive, vital because it prevents us from breathing in the evil odors of swamps and corpses, fruitless because it plays little role in the acquisition of knowledge.

For animals, the nose was "a universal organ of feeling," wrote Buffon, a kind of eye that sees not only where objects are but where they have been. For man, however, guided by judgment and reason, it is the lowliest sense; our lack of olfactory awareness elevates us above other animals. "Man must think more than he must crave," Buffon wrote, "and the animal must crave more than he must think."

Darwin believed smell only a distant vestige of what had been a predominant characteristic in a far-off ancestor and had degenerated because of little use. Whatever biological value smell may once have had faded when we assumed an erect posture and turned our nose away from the earth; now we need it scarcely more than we need long, fingerlike toes.

What the nose knew was hardly worth knowing.

Or naming. We are eye- and ear-minded creatures, hoarding our vivid vocabulary for the wham and the dazzle, not the musty, sweetish aroma of earth before rain. Sensing anything is partly a matter of articulation; but the features of smell defy our attempts to bring them under the net of language. In describing odors, we fall back on analogy, on metaphors from other senses, or simply refer back to the source. Things smell like smoke, like fruit, like jasmine or pine, like caramel, raspberry, vanilla. Jasmine smells floral. Yeast smells, well, yeasty. Even Nabokov, that master of language, found the subtle perfume of butterfly wings difficult to describe.

Plato postulated that odors lacked names because they did not consist of a definite number of simple types. Linnaeus — a compulsive classifier who gave us such a beautiful tool for ordering plants and animals — aimed to improve on this situation in 1752 by organizing odors into seven classes:

aromatic
fragrant
ambrosial (musky)
alliaceous (garlicky)

hircine (goaty)
repulsive
nauseating

Hendrik Zwaardemaker, a Dutch physiologist, slightly revised and expanded this system in the late nineteenth century, subdividing the aromatic into five categories, from camphoraceous to almond, redefining the fragrant as the "floral" and "balsamic," and adding the "ethereal" and the "empyreumatic."

A blunt system something like Zwaardemaker's is still in use. And still we have no good scientific method for measuring odors, no linear scale, as we have for wavelengths of light and frequencies of sound. I imagine Lucy possesses a taxonomy and scale far more exact than ours.

In 1752, the same year Linnaeus caught smells in his matrix of classification, a Dutch physician proposed that odors arose from a "fluidum," a "guiding spirit," oily and indestructible: "Exhaled by roses on a spring evening, the odor returns to the rosebush with the morning dew." A lovely idea — but fanciful. Not until the eve of the nineteenth century did science understand that odorants are made of chemicals. Mostly small and pithy, they vary widely in composition and three-dimensional shape. A single jasmine flower or fern may emit dozens of different odor molecules in highly specific mixtures.

The genius of the olfactory system, whether in mouse or man, lies in its ability to accommodate the random component mixtures of this chemical universe. The multiple odor molecules emitted by a sun-warmed fern are carried in eddies of inspired air to the roof of my nasal cavities, where, at the top of the nose, just below eye height, they land on a pair of mucus-bathed patches of epithelial tissue. Here they encounter the specialized neurons known as olfactory receptor cells. In the membranes of these cells are embedded protein receptors that recognize and bind to the odorant molecules (protein binding to protein), prompting the nerve to fire a message up to the olfactory bulb, a sort of relay station, where they're decoded and sent on to the higher olfactory regions of the brain.

The trouble we have articulating and classifying the world of smells may be rooted in the particulars of our anatomy. In the human brain, olfactory neurons have relatively few direct connections with the neocortex, the thinking cap of gray matter that houses our center of language. The links are tight, however, with the limbic region of the brain, the seat of emotions and memory, sexuality, appetite, body temperature, and much more. This may explain why odors possess the power to suggest, evoke, frighten, arouse, to make one mystical, stir one's passions, wake the memory of old romances, trouble the brain, stain the imagination.

When I was ten, I got lost in the woods of Vermont. I wandered for hours, looking for signs of a trail. Exhausted, I found a bed of ferns in a hot lozenge of sunlight and collapsed in a crying heap. I eventually wove my way out of the woods, but to this day the scent of sun-warmed ferns unsettles me, for in it I smell fear.

The odor of creosote pleases my father, who finds it a deep groove going back to pleasant boyhood work projects in a summer camp that offered cool green escape from the heat and bluster of New York City. The aroma of my grandfather's pipe and the fragrance of graham crackers dipped in milk slightly soured goes directly to the corner of my brain devoted to childhood and evokes an earlier world. Even deeper goes the sweet foresty smell of pine and moss. While the strong, exotic odors of orchids, cardamom, and saffron excite me by their newness, the smell of pine touches some atavistic nerve in me, some blood-remembering beyond the specifics of time and place. In it I smell home, as if my Russian and German ancestors had passed along in their genes their home roots in landscapes of dark, coniferous woods.

Once or twice a year, a boy is born with small genitalia, sterility, and no sense of smell. This disorder, known as Kallmann's syndrome, is a rare genetic disease, linked to a missing gene on the short arm of the X chromosome. The gene normally produces a protein that helps nerve cells find their way to their proper locations. Cells that make sex hormones originate in the same part of the embryonic brain as olfactory cells. In people with Kallmann's syndrome, the cells fail to migrate, and patients are born without the proper neurons to make sex hormones and without an olfactory bulb.

Imagine living without smell. When my grandmother began to suffer from Alzheimer's disease, her sense of smell diminished and, with it, the savor of coffee, chocolate, garlic, wine, life. It's a loss I fear. Imagine a pine forest without its resinous scent, fresh baked bread without its warm, yeasty odor. One man who lost his sense of smell said it was, in ways, worse than being struck blind. "You *smell* people, you *smell* books, you *smell* the city, you *smell* the spring — maybe not consciously, but as a rich unconscious background to everything else."

The universal ability to sense the chemical richness of the world likely arose billions of years ago, enabling the earliest cells to respond to substances without taking them in. Throughout the long pull of evolution, the basic structures for chemical sensing have changed little. The olfactory cell on an antenna of a moth matches that in the olfactory rosette of a cuttlefish, in Lucy's nasal cavity and my own. In all animals, the cell is "bipolar," extending a thin dendrite out to the environment and projecting an axon at the other end straight up toward the brain.

Then why such a gulf between Lucy's abilities and mine?

The secret lies in the deep molecular details of odor reception. Parsing them is a tricky business. The nature of sensory receptors for vision were uncovered long ago, but the proteins that act as smell receptors and the genes that make them have been nearly as elusive as words to capture the sensations they spark.

The neurobiologists Linda Buck and Richard Axel finally ferreted them out in the early 1990s in a study of rats. What they found was not a few key genes, as they had expected, but a huge family of a thousand genes never before seen, all different but closely related. (The similarity of gene to gene suggested that the rat's olfactory repertoire is the result of lots of gene duplications. So, too, it's likely, is our own.)

These genes spell out versions of serpentine proteins with the ungainly name of G-protein-linked receptors. Members of a family of proteins sharing a snakelike motif, the long, thin receptors thread through a cell membrane, crisscrossing it seven times. They are part of an ancient, versatile system of signaling, molecule to molecule. From their spot, embedded in the cell membrane, the receptors pass information from outside the cell to a class of molecules inside the cell, the G-proteins. It is the serpentine G-linked receptors and their complementary G-proteins that underlie the working of all kinds of biological exchange, that allow yeast cells to recognize mating signals, the cells of our gonads to respond to our reproductive hormones, the cells in our eyes to respond to light, our taste cells to tastes — at least to bitter ones — our brain cells to lock on to opiates. In smell, the G-linked receptors act as go-betweens in the conversation between the odorant itself and the body's molecules that pass the message along a nerve cell's axon and through the nervous system. A single receptor can recognize different odorants, and different odorants can latch on to different arrays of receptors, so the limited alphabet of a thousand or so receptors we humans possess can distinguish some ten thousand odors.

With the codes gleaned from the study by Axel and Buck, scientists fished for odor receptor genes in libraries of DNA from other species and turned up similar genes in salamanders, mice, dogs, and humans. The fruit fly also possesses a family of such G-protein-linked receptor genes for smell, but the fly's genes bear little base-by-base resemblance to our own. Its genes seem unique; no counterparts have yet turned up in other species, raising the philosophically interesting possibility that the fly's odor world may be altogether different from our own.

The really big surprise of Axel and Buck's study was the size of the gene family, the largest so far identified in mammals. One percent of the rat genome — one out of every hundred genes — is devoted to the detection of odors, which suggests that smell plays an even bigger role in mammalian reproduction and survival than we had imagined.

Humans have roughly half the number of olfactory receptor genes found in a rodent, an estimated 500 to 750, scattered across most of our forty-six chromosomes. With these we can detect thousands of odors — frightened skunk, fragrant lily, fresh lime, peppermint, lavender, turpentine, jasmine, vinegar, fine wine. That's the good news. The bad news is that almost three quarters of these genes are "pseudogenes," so hobbled by mutations that they've ceased to function, which may help to explain the relative mediocrity of my equipment compared with Lucy's. Dogs, marmosets, mice, rats, all appear to have their olfactory genes intact.

What's more, with the passing of time since we diverged from our ape cousins, human smell genes seem to have rapidly deteriorated. Our not-so-distant primate ancestors had far more sophisticated olfactory gene machinery. A study comparing the olfactory receptor genes among ten primate species revealed 19 percent pseudogenes in the baboon; 48 percent in the chimpanzee; and 50 percent in the gorilla, the primate species most closely related to us.

I mourn what I am missing. I'd love on occasion to lift my nostrils and sniff the rich, subtle world sensed by my older animal ancestors, to gather in the chemical babble of life with the acuity of a marmoset or a vulture or, for that matter, a fruit fly.

What if we could mend those pseudogenes, remind them of their original purpose, and transform the human nose into the incorruptible, hard-working organ of other species? Specialists in smell, perfumers and the like, can find well-hidden distinctions in scents. If we took this talent a few steps further, learned the fractionary smelling and dissecting of scents practiced by other animals, we might decipher complicated odors by composition and proportion, and find in subtle smells not a great blank but Big Meaning. We might reorder the hierarchy of old philosophers, so that what mattered first in the world was the swirling up of ecstatic aromatic plumes, airy, delightful, evil. We might devise a sophisticated taxonomy of smells and enrich our vocabulary to tease apart meaning in the crowded, beloved vapors of coffee, tea, chocolate, cedar, mace, anise, eucalyptus, honeysuckle.

Or would we be overwhelmed with the sudden swelling of scent, set swaying by unnameable odors? Would an indescribable hazy chaos reign in our biological house, as if each musician in a thousand-member orchestra were playing a different loud melody in odorous cacophony? We might suffer the effects of our higher neural anatomy, the tight cranial web of smell, emotion, memory, so that we were ever reeling under a constant, heavy swing of feeling.

Still, I wouldn't mind taking the risk. The neurologist Oliver Sacks told of a young medical student high on amphetamines who dreamed he was a dog during a night of fitful sleep. When he awoke, he could smell with all the exquisite faculties of a bloodhound. Suddenly the world was redolent, awash with particular odors. He discovered that he could find his way around his neighborhood, his city, by smell alone; that each of his friends and patients had a distinctive "olfactory physiognomy, a smell-face, far more vivid and evocative . . . than any sight face." All other sensations paled before smell. According to Sacks, the episode lasted only three weeks, and when it was over, the young man felt a mixture of loss and relief. "The storm of sensation had been exhausting," he told Sacks, but it had opened a world of "pure perception, rich, alive, self-sufficient, and full. I see now what we give up in being civilized and human."

History is full of unusual cases of olfactory sensitivity. "I like very much to be surrounded with good smells," wrote Montaigne, "and I hate bad ones beyond measure, and detect them further off than anyone." Napoleon claimed that he could recognize the land of his birth with his eyes closed, solely by its odor. (Certainly he could smell his Josephine and recall her odor with lust.)

So, too, are the annals of medicine peppered with such cases: a toddler capable of differentiating odors from various parts of the body; a man so sensitive to chemical odors that he could detect them wafting from a drugstore; the biologist who conducted a series of brilliant experiments on the long-range sex attraction of Saturniid moths and found that he could detect streaks of moth odor in the street near his house; a German physiologist, A. Bethe, who claimed to smell changes in people caused by subtle emotional excitement, menstruation, disease.

Hyperosmia, the condition of abnormally keen smell perception, is thought to occur fairly often among people with certain neuroses and psychoses. One report from the psychiatric ward of Bellevue Hospital listed forty cases of psychoses with olfactory hallucinations. In the *New York State Journal of Medicine*, Harry Wiener, a physician, wrote that "some of these case reports sound uncomfortably like Bethe's self-description of his olfactory perceptions." Wiener wondered whether certain supposed olfactory hallucinations were not descriptions of reality by patients "whose olfactory . . . acuity far exceeds that of their keepers."

Like so many of nature's workings apart from the beaten path, the experience of smell prodigies hints at a hidden possibility: that despite our failing equipment, *H. sapiens* may possess the biological means to detect a wider range of chemical messages than we suppose, even if they elude conscious perception by most of us.

"I have found that the idea of chemical messages passing between me and thee, without either of us being aware of them, is a very hard one to swallow," wrote Wiener in 1966, "particularly for men concentrating on the basic sciences."

The word "pheromone" (from the Greek *pherein*, to carry, and *hormon,* to urge on) had been coined only seven years earlier to describe a chemical signal passed between members of the same species — the sexual attractants of the butterfly, the warning substance of the minnow, the territory-marking substances of the Carnivora. Ferried in currents of water or air, pheromones send messages that cause alarm in a toad, sexual arousal in a bull, aggression in a boar. Mammals use them to mark territory and advertise sexual readiness. When an ovulating female boar catches a whiff of a male's urine, she lifts her haunches in readiness for mating. Pheromones in the urine of a female mouse can slow down the estrous cycle of another female, while those in the urine of a male mouse can accelerate it.

The first pheromone identified was bombykol, the sex attractant of the silkworm moth. It took half a million female silkworm pheromone glands and thirty years of chemical analysis to break the chemical code of bombykol. Each molecule, made of sixteen atoms of carbon, thirty of hydrogen, and one of oxygen, has two characteristic twists in its three-dimensional structure. Without these twists, the compound would arouse the male silkworm moth scarcely more than plain air.

The chemical nature of many pheromones is still largely a mystery. They tend to be active in extremely small concentrations and can cut through the chemical babble of the air to relay messages with laserlike efficiency. The amount of bombykol possessed by a single female moth suffices, theoretically, to drive more than a billion males into a haze of ardor. Lately, biologists have measured the sensitivity of the mouse neurons that respond to pheromones and found that they are a thousand to ten thousand times more sensitive than are standard olfactory neurons tuned to normal "perceptible" smells.

By means of pheromones, termites, bees, wasps, and ants act in unison, build artful dwellings, divide their labors, attack others, and defend their own. The "spirit of the hive" arises from their influence. In forming swarms, adult male locusts secrete pheromones that accelerate the growth of young locusts so that they can join the melee. Fire ants blaze scent trails to food for the colony to follow by touching their stingers to the ground intermittently, according to Edward O. Wilson, like a moving pen dispensing ink. A single milligram of the leaf-cutter ant's trail chemical could lead a small column of ants three times around the world. If a worker ant is disturbed, it discharges a chemical alarm, says Wilson, "an urgent tocsin that can propel a whole colony into violent and instant action." I'm happy to

learn that alarm pheromones don't travel as far or linger as long as sex attractants, which are enduring and present everywhere.

The possibility that humans are involved in this sort of social communication got a boost in 1998 from a study by the biopsychologist Martha
McClintock, suggesting evidence of at least two human pheromones with
measurable effects on the bodies of others. Twenty-five years earlier, Mc-
Clintock had noted that the menstrual cycles of young women living at
close quarters tended to converge over time, a kind of spontaneous synchrony that occurs in sheep, pigs, lemurs. In her new study, she found that
underarm sweat taken from women in a late follicular phase of their menstrual cycle shortened the menstrual cycles of recipient women. Odors
taken on the day that donor women ovulated had the opposite effect,
lengthening the recipients' cycles. McClintock proposed that pheromones
were responsible.

It was once thought that we had inherited only a pitiful archaic vestige
of the organ needed to perceive chemical messages, nothing like the robust
vomeronasal organ owned by a dog or even a monkey, a pair of small
sacs just behind the nostrils and above the hard palate. The human
vomeronasal organ showed up in the fetus, then seemed to atrophy or disappear altogether. But scientists taking a closer look have found that we are
in fact equipped with a small tubular vomeronasal organ with tiny open-
ings into the nasal cavity, and that we, too, possess at least one gene that
codes for a pheromone receptor — but whether gene or organ works in any
meaningful way remains an open question.

Likewise, we still don't have the ghost of an idea whether "odorless"
chemical signals from one human can be detected by another. Nor do we
know the full range of their possible effects, immediate or delayed. But it
seems clear that we're emitting unaware some intimate essence of ourselves
connected in mysterious ways with the essence of family.

How odd to think that members of our tribe — we who can whisper
an overture, sing out the blues, gesture our annoyance, scribble a love note,
beseech and entreat with fine manipulations of lip and jaw — may still
pass important messages about the nature of relations — self, family, mate,
other — by means of body moisture. On the other hand, most animals engage in this kind of conversation. Why should we suspect we're a wholly
special case?

12

TWO BROTHERS, EIGHT COUSINS

MORE THAN A HUNDRED YEARS after Darwin observed his son's sensual awareness of his mother, scientists discovered that infants as young as six days old turn to face a pad that has been worn near the mother's breast rather than a clean control pad or one from an unfamiliar nursing mother. Even a baby three days old can pick out her mother's breast odor and will make suckling movements and stop flailing her arms when her nose touches her mother's breast pad. When exposed to breast pads of other lactating women, she will turn away or ignore them.

This may be so, but stories abound of kin recognition gone awry in human life, of son not knowing mother, or father son, or brother sister. Oedipus, doomed to fulfill the terrible prophecy for his life, failed to recognize his father, murdered him, and married his mother. Isaac, unable to distinguish between his two sons, allowed Jacob to cheat his older brother, Esau, out of his birthright.

Science, on the other hand, tells one tale after another of animals with the uncanny ability to spot kin at all stages of life. A Mexican free-tailed bat can find her young in a maternity cave of millions of bats. Rats, mice, frogs, and insects can distinguish kin from unrelated animals even when they've never seen them before. Sweat bees guarding their hive will admit unfamiliar kin but toss out unrelated interlopers. Even the tadpole-like larvae of the sea squirt, *Botryllus schlosseri,* knows its relatives and settles with its siblings in colonies on the sea floor.

And while these creatures cannot conceptualize such categories as cousin or grandfather, they can sort their relatives to an astounding degree, not just parents and offspring, but half-siblings, cousins, grandparents, and uncles, and often use the degree of relationship to forge bonds or set behavior. A Japanese quail can tell a first cousin from a third cousin. Some kinds of tadpoles are picky enough to prefer schooling with full siblings rather than half-siblings. The blind goby fish, a little cannibal I first met among the snail-ridden grasses of a Delaware salt marsh, will eat alien young of its species but never its own progeny.

Salamanders, too, practice a form of relative cannibalism, choosing to devour those young most distantly related to themselves. And here's a surprise: their calculations on relationship affect not only behavior but morphology. David Pfennig, a biologist at the University of North Carolina, has found that a tiger salamander raised with a group of siblings develops into a small invertebrate-eating creature. But if the salamander is reared in a mixed brood, with nonrelatives, it grows into a larger, cannibalistic beast, with a broad snout and long curved teeth designed for catching and ingesting other tiger salamanders, usually distant relatives. Tadpoles of spadefoot toads undergo similar morphological changes in response to the presence or absence of kin.

Even plants show sensitivity to relations. Wild garlic surrounded by genetically identical neighbors grows bigger and better than garlic surrounded by unrelated plants. Plantains and pokeweed also flourish when potted with kin.

My sister Kim is a schoolteacher. In keeping with the spirit of her profession, she once brought home from the classroom a chicken egg in an incubator to care for over vacation. As luck would have it, the egg hatched on her watch. The chick took my sister as family and followed her everywhere — the same imprinting behavior made famous by Konrad Lorenz's geese.

In the natural world, kin may be recognized as those in the right place at the right time, be they caretakers or siblings. This familiarity technique — "kin is who I see first" or "kin is where the home is" — occasionally backfires, even in natural settings. A warbler or wagtail may fail to eject from its nest the conspicuously different chicks of such parasitic species as cowbirds and cuckoos. I've seen film footage of a warbler mother frantically stuffing worm after worm into the mouth of a gargantuan cowbird chick bigger than she. Bank swallows also feed whatever nestlings they find in their burrows.

More reliable clues to family may be observable and distinctive — a single physical trait or a constellation of traits, such as body color, facial features, vocal signature, hair color, scent — shared to a greater or lesser extent by all family members. The tadpole of the dark, warty toad *Bufo americanus,* common in the damp, secret spots of my garden, may grasp its sense of family with the help of a chemical made by the mother and delivered to the egg before it hatches — either in the cytoplasm of the egg itself or in the jelly that surrounds it — which acts as a chemical family trademark. So, too, the English plantain likely uses chemical cues released by roots to tell kin from nonkin. The sweat bee seems to carry in its head the odor of its colony, learned by sniffing its nest mates. A sweat bee guarding a hive may be able to assess the degree of kinship by comparing the signature of a newcomer to the known family or colony cue — a chemical signal that overrides all other smells emitted by the bees or present in the hive. It will bar or admit the stranger according to its degree of relatedness. Primates tend to spot family through visual cues. Chimpanzees faced with the task of linking portraits of unfamiliar chimps can match the faces of mothers and their offspring.

Generally, we humans recognize our relatives because we grow up with them or are told who they are, because we have shared names and detailed genealogies. These strategies may have evolved precisely as mechanisms for identifying kin, David Pfennig told me. "In essence, we may use words as signatures of kinship, just as other organisms rely on scent and sound to identify relatives." But science suggests that we share with other organisms a sensitivity to quieter, more subtle family cues.

In a shoebox I keep in the corner of a closet is a silk scarf that once belonged to my mother. Though she died twenty years ago, I still feel orphaned by her loss, still miss her fiercely. I know I'll never hear her voice again, a voice that radiated warmth, the acoustics of unqualified love. But in the soft silk weave of this scarf, I imagine I can faintly smell the particular, indescribable mixture of her being, and in it I find a source of enormous comfort.

I may well have learned my mother's odor before birth, as my daughters learned my scent. We humans are remarkably well scented, better endowed with odor glands than any other primate; they cover the body from head to toe, residing in nooks and crannies of face, scalp, upper lip, nipples, penis, scrotum, pubis, even eyelids and ear canals. They're especially abundant in our axilla, or underarms, organs with tufts of hair perfectly tailored to wick odors out into the world. Many of these odors contain steroids, compounds that coordinate sexual reproduction in much of the animal world. Some smell like musk, a sex attractant made by male Himalayan deer in a small pouch near the penis during the rutting period. We consider the odor of musk warm and comforting, and seek it as an essential ingredient of expensive perfumes. So, too, civet, a sex attractant originally isolated from the glands near the anus of the African civet cat.

Our ability to navigate the world by smell may be many orders of magnitude inferior to that of a hound, but the human nose has remarkable powers of discrimination when it comes to sniffing kin. A blindfolded mother who has just given birth can pick out her infant by smell. This is true even if she has delivered by Caesarean section and has had little contact with the baby before the test. An infant's other relatives can recognize the child by smell, too, but they require a longer period of exposure: aunts, three quarters of an hour; grandmothers, twice that.

I'm certain I could sniff out my young daughters in a crowd, and they, me.

In 1802, the physician Pierre-Jean-George Cabanis wrote:

> It is quite obvious that each species, and even each individual, emits a particular odor. It gathers around him in a cloud of animal vapors, constantly refreshed by the life he leads: and when that individual moves, he leaves behind him particles that enable animals of his own species, or of another that has a keen sense of smell, to track him.

Cabanis was right. Each of us is marked by a scent signature that is stable over time (unlike visual appearance and voice, which change from birth through old age) and heraldic, too, with peculiar and particular elements common to the family.

In the 1980s, the psychologist Peter Hepper set out to tease apart the possible sources of distinctive human body scent with a series of experiments on trained tracker dogs. Was it diet or another aspect of environment? Or was it genes?

First, Hepper presented his dogs with nonidentical human twins, a few months old, who lived in the same house and were fed the same diet.

The dogs instantly smelled the difference between the twins. They had more difficulty with identical twins, even those who lived apart and ate different diets. And with identical twins living in the same home and eating the same food, the dogs failed completely to distinguish between the two.

Apparently genes were a key. But which genes?

A decade earlier, in the mouse-breeding rooms at Memorial Sloan-Kettering Cancer Center in New York, researchers unknowingly got a hint about the genetic source of individual body odor. The researchers noted peculiar social behavior in a special colony of so-called congenic mice, bred to be genetically identical except for a string of linked genes called the major histocompatibility complex, or MHC. The MHC genes are best known for their role in organ transplants; they help the body to distinguish between self and nonself, telling it to reject infectious agents or foreign tissue from unrelated individuals. MHC genes vary a lot within a species but remain relatively constant in families.

This was the strange behavior: the congenic strains of mice at Sloan-Kettering seemed to sniff out the minute genetic divergences among their cage mates and to select their company accordingly. It was difficult to get a pair to mate if the two had similar MHC genes. Later, it was found that female mice are apt to choose as mates those males with an MHC different from their own and to nest communally with females possessing a similar complex. The composition of MHC genes, it turns out, does affect body odor. Because related individuals tend to have similar MHC genes, they also have similar body smells. So mice can literally smell their family.

Virtually every vertebrate from wood frog to wolf carries the MHC set of genes, though it varies widely in size and shape from species to species. A chicken has a set of only nineteen such genes; humans, a set of four hundred or so, all on Chromosome 6. In humans the set is called the human leukocyte antigen, or HLA system, and is by far the most variable part of the human genome. Some parts of the HLA are a kind of genetic hotspot, a place in the genome where mutations occur frequently. As a result, there is a huge number of alternative forms for each of the genes, as many as a hundred. Each person carries a unique combination. For two people to have identical MHC genes is rare, except in cases of twins. That's why it's often difficult to find organ donors outside one's family. The amazing variability among MHC genes labels each body cell as unique to its owner and, hence, each individual as different from any other creature that has ever lived.

Mice are not alone in their ability to detect the subtle genetic differences in MHC genes. Young arctic char, relatives of salmon, prefer water scented by a sibling with similar MHC genes than by a sibling with a more divergent complex. Even sea squirt larvae sense these minute genetic differences and settle near other larvae with a like complex. Remarkably, we humans can smell the different body odors of mice that vary only in their MHC genes. And here's news to lift the nostrils: we may even be able to detect such divergences in our own kind.

In parts of the Dakota, Montana, and Canadian prairies live small, isolated agricultural communities of a religious group known as the Hutterite Brethren. Each colony, called a *Bruderhof* ("dwelling place of brothers"), imagines itself a kind of Noah's Ark — closed, self-sufficient, an oasis in a Godforsaken world. Hutterites have a saying: "Blood is not water." Members marry among their own and have large families. The group is remarkably homogeneous — indeed, inbred — all thirty-five thousand of the members being descendants of four hundred Anabaptists, who came to the United States from Europe in the 1870s to escape religious persecution. These four hundred, in turn, were descended from only ninety individuals. Thus, the Hutterites as a group share more HLA genes than do members of the population at large. But when Carole Ober, a geneticist at the University of Chicago, studied the DNA of Hutterite couples, she found that the HLA genes of wives and husbands differed more than one might expect from chance. This suggested to Ober that the Hutterites somehow manage to choose partners with HLA genes least like their own, perhaps relying on smell or pheromones as a cue.

There's good biological reason to select a mate with a different set of MHC genes. Scientists speculate that such pairs have offspring more resistant to disease. The MHC genes that label each of us as individuals also drive the immune system. The genes code for proteins that are carried as markers on the surface of our cells. From these markers, the immune system can tell whether a cell belongs to us or to an invader and whether it's healthy or infected with a virus. Having variability in MHC genes means that we can recognize all kinds of infectious organisms and respond to them effectively by making a range of antibodies. Here's a beautiful example of nature's parsimony. A single site on a single chromosome controls both our ability to fight disease and our characteristic odor, which marks each of us as unique and as a member of a family. Once the immune system evolved, the theory goes, it offered a means to distinguish not just self from nonself, but similar nonself (kin) from dissimilar nonself (nonkin).

Picking mates with unlike MHC genes has other advantages. "Nature abhors perpetual self-fertilization," Darwin once wrote, noting that some plants that fertilize their own flowers produce fewer seeds and stunted seedlings. For many small, short-lived arthropod species, this is not necessarily true. (Certain ants, mites, and button beetles do very well by inbreeding, making a habit of brother-sister incest. One antlike parasitoid of beetles carries the sin to extremes. The female as a matter of course mates with her son, then mates with a grandson born of a daughter begot from her previous incestuous union.) For larger, longer-lived creatures, however, such as birds, fish, humans, trees, even certain insects, inbreeding has hidden genetic traps. Nature must know this; the taboo against incest is not peculiar to our tribe but is observed by many species. Inbreeding can bring together rare dangerous recessive genes, which are normally diluted by mating outside the family. As any dog breeder can tell you, after many generations of inbreeding, purebred animals often suffer such serious disabilities as hip dysplasia, eye trouble, behavioral disorders. A harmful recessive gene doesn't wreak havoc unless it exists in two copies. Then it can emerge from hiding and cause great damage. Witness those four unfortunate children with true microcephaly born of first cousins, or the hemophilia that plagued Europe's royal families in the 1800s.

Just how many deleterious or "lethal" genes are floating about in the gene pools of living things is a matter of debate. For humans, one study reckoned 1.6 per person. That means that any pair of random people in the population aren't likely to have matching lethals, as Richard Dawkins explained in his brilliant book *The Selfish Gene*. Close relatives, however, are.

> If you have 100 rare genes, approximately 50 of them are in the body of any one of your brothers or sisters. . . . However rare my lethal recessives may be in the population at large, and however rare my sister's . . . there is a disquietingly high chance that hers and mine are the same. . . . If I mate with my sister, one in eight of our offspring will be born dead or will die young.

Grim insights into the effects of inbreeding on wildlife arrived lately from the Åland Islands in southwest Finland, home to a dwindling population of the Glanville fritillary, a butterfly. The fritillary's dry-meadow island habitat has been diminished and fragmented, isolating small populations. The result has been inbreeding and loss of genetic diversity, which has caused the species serious injury. With each breeding season, fewer eggs hatch, fewer larvae survive, and female adults die earlier, factors that are slowly driving the butterfly to extinction. Scientists speculate that the Glanville fritillary may be a kind of canary in the coal mine, alerting us to the genetic effects of habitat loss or fragmentation among many threatened species.

Inbreeding is indeed an evil for all sorts of organisms to avoid. But for most of us, the ability to recognize self, family, and other has a hidden, possibly more critical, purpose. While studying wasp behavior in the 1960s, the evolutionary biologist William D. Hamilton noted that a female wasp was more likely to sacrifice herself for close relatives than for distant ones. He concluded that the altruistic wasp was giving up her life to ensure the survival of copies of her genes in the bodies of her relatives. Flesh perishes; genes live on. If a wasp dies in order to save ten close relatives, a single copy of her gene may be lost, but a larger number of copies of the same gene is saved. (This may also explain the self-sacrificial behavior of worker bees, plovers, and naked mole rats.) Evolution makes no distinction between copies of genes possessed by direct descendants, such as offspring, and those of siblings. What counts in "inclusive fitness," as Hamilton called it, is not just the survival of the individual, but the survival of that individual's complement of genes, taking into account children, brothers, sisters, nieces, nephews, and so on. Natural selection favors organisms that help near relatives. In the words of the British geneticist J. B. S. Haldane, "I would lay down my life for two brothers or eight cousins."

Is it any accident that the word "family" derives from the Latin *famulus*, meaning servant of the household?

Lewis Thomas once pointed out that Haldane's mathematical sequence invites expansion. In terms of genetics, it makes sense for me to give my life for the sake of two siblings or two children, four nephews or four aunts or four grandparents, eight cousins or eight great-grandchildren, thirty-two second cousins, one hundred and twenty-eight third cousins, and so on. In cold-potato mathematical terms, the tight genetic connections of family end with third cousins. My chances of sharing a particular gene with relatives that distant are much the same as with any random individual. But still, we are related. Even two random individuals don't have to go back far in their family trees to meet a common ancestor. Scientists deciphering the statistics of genealogical trees have shown that my neighbor and I probably shared an ancestor within the past thirty generations; that, in fact, most people are related to anyone who lived before the thirteenth century — say, Alfred the Great, Charles the Bald, Charlemagne. All of us alive today are more or less distant relatives.

As Thomas wrote, the matter does not end with our species. Imagine Haldane's progression going even farther outward — and, along with it, genetic responsibility — out into remote corners of relation, even beyond the bounds of our own species. Cousins, after all, are defined by ancestry, whether in the family Kennedy or Canidae. While it's true that wolf and deer are hardwired in ways peculiar to their species, so that carnivore goes after herbivore, it's also true that they are, in a real, measurable sense, cousins.

Here's something I hope to remember: unique as we are, specified by highly individual little chemical labels, which help protect us from the perils of inbreeding and disease, we are also, all of us, relatives, bound by the same filament, transmuted at relatively few removes. We have not a single share in the corporate Earth but are invested all over the place, in remote cousins from whom we are separated only by a matter of degree, measured in small base-pair differences in DNA.

13

A KIND OF REMEMBRANCE

SOME THINGS REMEMBERED are remembered forever, imprinted, burned into the brain like the eye of that injured bird that still surfaces in my memory to haunt me. It's funny how the mind will sequester certain memories, treasure them in some dark recess, then suddenly allow them to flare up again like the fever of malaria; funny how it will graft two memories, bring them into juxtaposition, as though by associating them it wished to produce a third.

I have a poor memory generally, full of holes, the vivid recollections like occasional patches of sunlight piercing dark swaths of unremembered greenery. Few events do I recall in detail, with the neat clarity of something seen through the wrong end of binoculars. Memory often works better when emotion or the perception of danger figures into the mix; something about the chemistry of fear or feeling more deeply imprints the recollection.

When I was eight — or seven or maybe ten — I met a dare from two friends to join a group for a midnight swim in the forbidden waters of a neighbor's pool. The neighbor was known for his dislike of children, his cold, even cruel demeanor. The night was warm and thick with the fragrance of honeysuckle. I remember stealing out of the house and meeting up with my friends, the four of us, or maybe five, pressing through the dark woods to my neighbor's house. I remember the squeak of the pool gate, the tepid water reeking of chlorine, the muffled laughs and splashes. Then the slam of a screen door, the sound of footsteps, a rising shadow. In a burst of panic, I scrambled out of the pool and over the fence, catching my leg on a spike of iron, or maybe it was wood, thudding down hard on my hands and knees on the other side, fleeing toward home, aware only of the damp denim of my cutoffs chafing my legs as I ran and a sticky wetness oozing down from my knee and pooling in my shoe.

I didn't reveal to my parents the thick slice in my leg for fear of drawing their scolding attention. The wound became infected, festered for a few days, then healed to an ugly white scar.

That the thin wires of my brain can hold on to any memory of this event strikes me as marvelous, especially when I stop to realize that in the three decades or so since its occurrence, every molecule in my mind has shifted away to become part of something else.

The secret to storing memories lies in the links, the synapses, where a dendrite from one brain cell touches the axon of another. Experience alters the synapse to create memory. Among the many molecules believed essential to the change is the potent and ubiquitous CREB gene. Animals with mutant CREB genes can learn new things but cannot remember them. CREB may play a role in switching on other genes to change the shape and working of a synapse, helping to lodge long-lasting memory in the minds of virtually all animals, enabling the scrub jay to find its nut cache, allowing the chimpanzee to recollect the heraldic lip and tuft of its near relatives, giving Vladimir Nabokov the power to pluck from the haze of past days a vivid childhood glimpse of a Norse goddess butterfly.

I've read that something like half of all children recall facts and events in photographic detail. In all but a few, this eidetic memory disappears with puberty, giving way to a memory that filters, sorts, judges, overlooks — the wispy geometry of adult memory. This is necessarily so. Imagine if we did hold on to all recollections, like the famous Russian mnemonist Shereshevsky, whose limitless memory, hopelessly burdened with everything it had ever encountered, was a junk heap of impressions and images that crippled him, flooded and disturbed his mind so much that he ended up doing little but performing memory tricks in a music hall.

The normal brain is necessarily choosy and crabbed in its recollection. It has no bin for whole happenings, no single cell or string of cells that holds the full moment. Memory doesn't mirror events or rerun them like film; it re-creates them from choppy, fragmented impressions and quick perceptual slices: the warm summer wind, the liquid darkness, my cutoffs heavy and wet. Some details slip away, unrecorded, or are misplaced in the mind's nebulous depths. Others are masked or mingled with other impressions released from neighboring brain cells. Washed from my mind are the names of my companions on that forbidden swim, the precise instrument of my injury, whether I got in trouble at home, even my age at the time.

In my body, however, is another, highly accurate remembrance of the event. Fixed in the memory of my immune cells is the exact identity and nature of whatever microbes made their way from fence post to flesh during my flight over that fence. No vague or ambiguous recollections here, but a photographic recall of the particular subtleties of each microbial character, a cold, precise, permanent memory of other organisms.

The specific and long-lasting memory of the body's immune cells was noted by Ludwig Panum, a Danish physician who investigated an outbreak of measles in 1846 in the Faeroe Islands of the North Atlantic. Sixty-five years earlier, the islands had experienced a measles epidemic, but had since remained measles-free. Of the many aged people still living on the Faeroes who had been exposed to measles the first time around, Panum observed, none contracted the disease during the second epidemic.

The fact of immunity was no surprise. During the plague of Athens in 430 B.C., Thucydides remarked that the same man didn't sicken twice from the same disease. (The Latin word *immunis* has its roots in the ancient legal concept of exemption from service. Marcus Annaeus Lucanus, a Roman poet and a nephew of Seneca, first used the word in the context of disease in the first century to describe the resistance of the North African Psylli tribe to snakebite.)

What was striking in Ludwig Panum's observation was that long gap when the population remained free of the measles despite the presence of the virus. Apparently the bodies of those who had survived the first epidemic vividly recalled their encounter with the pathogen, even though they had gone for more than half a century without the slightest refresher.

There is no way to pack an immune system into an ice chest, no way to get a slice of the whole thing under a microscope. The system has about as many cells as the brain or liver, but is not nearly as tangible or well defined. It is more akin to an ecosystem than to an organ, an assembly of trillions of components. I think of it as an inner galaxy, illuminating the dark contours of my body with trillions of sparkling gypsylike cells roaming the blood and lymph, communicating through a legion of messengers, complicated, coordinated, fantastically skilled. Among these are some ancient primitive defensive elements, nearly universal in nature.

In the 1980s, when the biologist Michael Zasloff was operating on African clawed frogs, surgically removing their ovaries to study their oocytes, he noticed that the wounds of frogs just operated on didn't get infected, even though the surgical procedures weren't sterile and the water tanks, to which the frogs were returned after surgery, were heavily contaminated with microbes. Puzzled, Zasloff studied the skin of the wounded frogs and soon discovered that the cells were making powerful natural antibiotics, which he called magainins (from the Hebrew word for "shield").

A few years earlier, scientists had found similar natural antibiotics in giant cecropia silk moths. Called cecropins, the antibiotics proved to be potent killers of bacteria. They help insects fend off infection and protect them during metamorphosis, when the insect's guts and internal organs dissolve, releasing a flood of microbes inside the creature's body.

Zasloff first discovered a mammalian version of cecropins, called defensins, in and around the grazing injuries on the tongues of cows. These days such antimicrobial peptides are popping up everywhere in nature, from the venom of wolf spiders to the white blood cells of humans. Though these natural antibiotics vary from insect to mammal, they work in the same way, by punching holes in the membranes of invading bacteria, causing the cells to burst. Zasloff suspects that all moist surface linings of a mammal, from cheek to gut, make these antibiotics; they have been found in the cells lining the human trachea, lungs, urogenital tract, and mouth. They're even present in the cells covering the cornea of the eye, in the placenta, and throughout the cortex of the brain.

More than a century before Zasloff's work, another universal player in the primitive immune defenses of nearly all creatures was discovered by a Russian zoologist probing a starfish with a thorn. One day when his family had gone to the circus, Élie Metchnikoff remained alone with his microscope. While he was observing the "mobile" cells roaming the body of a transparent starfish larva, he wrote, "It struck me that similar cells might serve in the defense of the organism against intruders. . . . If my supposition was true, a splinter introduced into the body of a starfish larva, devoid of blood vessels or of a nervous system, should soon be surrounded by mobile cells . . . as in the man who runs a splinter into his finger."

When Metchnikoff plunged a rose thorn into his starfish larva, he saw through his lens, with excited delight, masses of amoebalike cells crawling toward the thorn and attempting to engulf it. These wandering cells, Metchnikoff proposed, enabled animals of all varieties to withstand attacks by microbes. He called the scavenging white blood cells phagocytes, from the Greek verb *phagein*, "to eat."

It was to his phagocytes that Loren Eiseley apologized when he lay on the pavement bleeding, the blood cells that, through his folly and lack of care, were dying like beached fish on the hot pavement. Crawling wonders, phagocytes were among the first immune cells to attend to my wounded knee. They recognize intruders in a general way, having learned over the long pull of evolution to spot potentially harmful microbes by reading certain generic "bad news" shapes, patterns of specific proteins or cell-wall components that are shared by large groups of bacteria, viruses, and fungi, and are so essential to their survival that they do not change. Phagocytes have protein receptors for spotting these particular patterns, and because such pattern-recognition receptors are encoded in our genomes, this kind of phagocytic defense is considered innate.

Phagocytes will engulf anything foreign — microbe, splinter, thorn — attempting to swallow the intruder in a bubble of membrane. But they have no memory. That ability belongs to lymphocytes, the evolutionarily younger stars in my immune galaxy, cells that target their enemies with great specificity and hold long grudges. Our bone marrow makes lymphocytes all through life, a million a second, sending them out to patrol the body's lymph and blood. They're protean in shape, sometimes spherical, sometimes flat like a planarian worm, the better to squeeze in and out of the liver, the spleen, blood vessels, and fluids that bathe the tissues.

So-called B lymphocytes, which mature in the bone marrow, make antibodies, Y-shaped proteins that recognize and attack antigens, the foreign molecules often carried by microbes. T lymphocytes — named for the thymus, the small organ in which they mature — kill microbes outright and send signals that boost the production of antibodies.

Immunity is largely a matter of proteins recognizing other proteins by shape. Both B cells and T cells carry protein receptors on their cell surface, tiny organs of touch designed to recognize foreign proteins. Antibodies and the receptors on T cells descend from an old family of proteins that probably showed up early in the evolution of multicellular life, when it first became necessary for specialized cells to recognize one another. Their job was to spot patterns of self-proteins; they eventually expanded their repertoire to include foreign molecules.

Lymphocytes love to explore. They roam the body, sensing and monitoring, asking one tactile question: is there, anywhere, a particular molecular configuration that matches mine? If a B cell bellies up to a foreign molecule floating about in the body and finds that it matches its receptor, a small spectacle takes place. The B cell enlarges and divides, reproducing furiously to create a colony of millions of identical cells, all prepared to synthesize, now and in the future, the particular antibody needed to attack and destroy the infiltrator — be it a measles virus, a cholera toxin, zebra dander, or even the cells of a transplanted liver. A primed B cell can pump out antibody molecules at a rate of ten million an hour.

Some microbes, such as viruses, don't wander about, exposed, in the fluids and tissues of the body; they hide inside body cells. My immune system has figured out a solution to this, too — with no help from me. T cells detect the presence of such microbes with the help of those specialists of self, MHC molecules. The job of the MHC molecules is this: to bind with bits of proteins inside a cell, carry them to the cell surface, and present them to T cells as little flags of health or infection.

Not long ago, X-ray crystallographers deciphered the shape of MHC molecules and found that each has a deep groove designed to neatly embrace fragments of proteins. If a cell is healthy, the fragments delivered by MHC molecules to the surface all come from self-proteins. But if the cell is infected, the MHC will deliver bits and pieces of the intruder along with its self-flag. When a T cell designed to kill infected cells spots such a configuration, it also starts dividing so that it can attack all such infected cells.

The lymph nodes containing growing, dividing immune cells are what we feel as swollen glands. The full flowering of immune cells against a specific infection often takes a week; hence the delay between the onset of measles and the immune response that cures it. After the pathogen has been destroyed, the body eliminates the excess immune cells through cell suicide. But certain members of the new colony remain — as nothing less than a precise, nearly permanent tissue of cellular memory. This explains the power of vaccines. The injection of a foreign molecule characteristic of an organism raises a population of immune cells and antibodies with expertise in detecting and destroying it.

For years the wisdom was that all antibodies were born the same. When they met an invader, they simply shaped themselves to it, like putty to a key. Now it's known that the system is much more marvelous. Each antibody comes into the world with a different, solitary idea. Vertebrate immune systems have hundreds of millions of different antibodies, each fashioned to spot an antigen of a slightly different shape, allowing us to recognize and respond to a fantastic range of microbial interlopers.

There's a conundrum here. Antibodies are proteins, and proteins are made by genes. How can there be hundreds of millions of different antibody molecules surveilling my body, an unfathomable font of shapes, when my genome has about a hundred thousand genes? How can so few genes make so many antibodies? To generate all the variety, it turns out, the body performs a remarkable trick. It creates new genes through a device known among some immunologists as the generator of diversity, or GOD.

We inherit only pieces of the genes that code for antibodies. As our lymphocytes develop in thymus and bone marrow, a sort of random assembly device shuffles the gene pieces, mixing and matching them in ever-changing combinations to form new configurations. They do so with the help of two so-called RAG proteins. These are DNA-chopping enzymes that cut the gene fragments in specific places so that they can be easily rearranged. The result is a library of antibodies that can "fit" an almost infinite variety of molecules on bacteria, viruses, nearly anything foreign that might challenge us. After the cutting and pasting, the fit may not be perfect. But — here's the genius — when a B cell first spots a matching antigen, many little mutations occur in its antibody gene segments at an astonishing rate, a million times greater than the rate of spontaneous mutations in other genes, creating still more diversity of possible shapes. Called somatic hypermutation, it's evolution at high speed.

When a lymphocyte divides, the antibody genes in the daughter cells hypermutate, creating minor variations, so that the next generation of cells likely contains a few with antibodies that more nearly match the antigen key and can lock in tighter. This, then, makes the cells divide even faster. And so it goes, until the body is churning out cells with antibodies perfectly shaped to match the invaders.

The point of all this wild elaboration of shape is so that our immune cells can recognize almost any molecule on Earth. So skilled are these cells, with such an acute sense of molecular touch, that they can differentiate between two proteins that vary by only a single amino acid — or even a single atom. If God is in the details, then my immune cells are infused with a divine presence. They're like little poets, carrying mirrors through the streets, ready to render an image of every created thing. And here's the clincher: these cells hold precise images not just of interlopers they've met, but also of ones they may meet, even those that have not yet evolved or been made in a laboratory. They hold memories of the past *and* a kind of telepathic memory of the future.

We possess all of these microbial profiles without ever being conscious of them. Imagine if we had to register the microbes as we do people: through subtle impressions of physical or psychological traits, of like or dislike, confidence or mistrust. Imagine if we had to carry in our heads a gallery of a hundred million different faces — distinguished by slight bumps and curves, like the quixotic forms in a Miró or a Kandinsky — and life depended on instantly telling one from the other. I tremble to think of what might happen if these millions of little profiles were added to the crush of conscious information that already presses in on my overstocked mind: the odd little memories, from the first six digits of π to the familiar long-legged stride of my neighbor Kate, not to mention the inconsequential snatches of conversation, quick visual glimpses, the thoughts that never pause long enough to be comprehended, or the ones that go round and round like toy trains never getting anywhere, the nagging worries of family, lovers, work. If my conscious memory were in charge of remembering the microbes and molecules my body once met or may meet, nothing would save me. I gladly give the task over to my T cells and antibodies, and depend on them to look out where I'm going.

No system is perfect. A few years ago, my father went to his doctor's office for a routine diagnostic test. The doctor had seen a shadow in an X-ray of my dad's lung that he wanted to investigate. The test involved the injection of dye into the bloodstream. While the needle delivered the dye, my dad watched — but only for a few seconds. Almost instantly, he began to itch all over, to feel tense and swollen; then he gasped for air and lost consciousness. Some component of the dye, possibly something as benign as iodine, had set off alarms in his body and produced an anaphylactic attack that undid his whole system, swelled his throat, shut his airways, made his heart stop cold.

By the time I reached the hospital, the threat of death had passed. The medics had managed to start my dad's heart before there was any lasting damage. Still, it took him months to fully recover.

Immune reaction, like cell suicide, is a hair-trigger affair. Here was a simple case of mistaken identity by the immune system, the same small blunder that makes the allergy sufferer sneeze in response to grass pollen or the asthmatic wheeze in the presence of dust mites. Allergy, asthma, and anaphylaxis are all part of one syndrome, a tendency of the immune system to overreact to common inhaled proteins known as allergens. The syndrome tends to run in families and is linked to certain MHC genes. In my dad's case, the immunologic overkill nearly cost him his life.

There are other risks. Pump out billions of immune cells every day, and chances are some rogues will slip into the mix — lymphocytes with receptors designed to home in on the body's own proteins. To prevent such cells from destroying their castle, the body orders them to commit suicide or go to sleep, a state of unresponsiveness called "anergy." But not all rogue cells obey. When such cells survive and thrive, they can cause such autoimmune diseases as multiple sclerosis, rheumatoid arthritis, juvenile diabetes, scleroderma, and myasthenia gravis, a disease in which outlaw immune cells make antibodies against proteins on skeletal muscle cells, and the affected individual becomes weak and has difficulty breathing.

In recent years has come intriguing news that some autoimmune diseases may not be instances of self against self but, rather, matters intimately linked to family.

A decade ago, researchers testing the blood of a young female laboratory technician were shocked to find traces of male DNA. The technician, it turned out, was pregnant with a baby boy, and the cells of the fetus were circulating in her bloodstream. Since then scientists have found that a child can literally get under a mother's skin and stay there, its cells lingering in her body for decades after pregnancy, making her a chimeric blend of herself and her offspring. Cells traffic in both directions during pregnancy; maternal cells also enter a child's body in the womb and remain there for years after birth. In fact, most of us harbor cells that belong not to our own bodies but to those of immediate family, mementos from our babies or from our mothers. These foreign cells may incite the immune system to launch attacks against a body's native tissue. Take scleroderma, a disease in which the immune system attacks the cells of the skin, lungs, and internal organs, causing hardening of tissues, joint pain, and even organ failure and death. Women with scleroderma have more fetal cells circulating in their blood decades after pregnancy (up to sixty-one in a tablespoon of blood) than do healthy mothers (who tend to harbor at most five).

That mothers often carry a double load of such stowaway cells, inherited both from their children and from their mothers, may explain why they suffer from more autoimmune disorders than do men or women who have never borne children. During childbirth hundreds of thousands of these cells are released into the mother's bloodstream. Most of them are swept away by the mother's immune system, but some may persist. Later, when the mother's immune cells spot the interlopers, they may begin an explosive campaign of defense.

Despite its foibles, acquired immunity is a sensational system — one possessed only by vertebrates. Pity the insects, lobsters, snails, and other invertebrates that make do with only phagocytes and other primitive innate defenses, unable to keep up with microbes that easily evolve to shift their shapes, with a memory worse than mine. For years these primitive defenses were dismissed as antique and archaic, unintelligent, indiscriminate, forgetful — stopgap measures that contained the infection only until the "real" immune response could take over with its highly specific weapons.

But then a decade ago, Charles Janeway, an immunologist at Yale, spoke up about a "dirty little secret" in his field. Inject an animal with an antigen alone, and chances are the antigen won't spark an acquired immune response. The lymphocytes will just sit there, even if their receptors perfectly match the antigen. Scientists need to add a potion of oils, salts, and bits of killed bacteria, including parts of cell walls — the kind of generic shapes recognized as bad news by the "primitive" tribe of phagocytes.

Apparently merely spotting an antigen isn't enough to make a lymphocyte react. Janeway suspects that our sophisticated immune cells need a second signal from another source, a tip on the nature of the antigen and what might be a suitable response, an interpretation of the news: innocuous or dangerous. And that, says Janeway, is where the older system has an edge — wisdom drawn from the deep past during the millions of years of coping with microbes. It is not the lymphocytes that read "foreign" in a microbe, Janeway suggests, but the older, wiser innate cells and proteins. It is they who recognize an antigen as dangerous, send a red alert to the naive lymphocytes, and get the whole immune response going.

So our elegant, intricate new system is built upon the old one and relies on its signals. Lately, researchers have found the messengers that act as go-betweens. They're Toll proteins, and they're present in defenses of organisms from tobacco plants to fruit flies, mice, and humans, suggesting that this red alert system has been around for a long while, perhaps since the dawn of multicellular life. (These Toll proteins are familiar from a different context: they help shape the body plan of developing fruit fly embryos. Here's another example of nature as bricoleur, reusing bits and pieces from other projects.)

From the evolutionary view, acquired immunity is a relatively new thing under the sun, having appeared only about half a billion years ago, in a sort of immunological big bang. The problem of how such a limber, sophisticated system came about has been for immunologists a kind of white whale. For one thing, parts of the system seem to have appeared *de novo* all of a piece. Antibody genes and something like our RAG proteins are present in the horned shark. But so far, no sign of a precursor has turned up in slightly more primitive organisms, say, the starfish, jawless hagfish, or lamprey. Where did the remarkable gene-shuffling proteins in our GOD come from?

One explanation stands on its head the notion of our special separateness and sanctity. The RAG proteins that help the body shuffle pieces of antibody genes act much like the DNA-chopping proteins of certain infectious microorganisms. The immunologist David Schatz and his colleagues at Yale have shown that RAG proteins can facilitate the hopping of genes, including RAG genes themselves, from one place to another, suggesting that they were once part of an ancient transposon, or jumping gene. Theory has it that 450 million years ago, a mobile bit of DNA — a jumping gene from a virus or bacterium — leaped from its chromosomal home and inserted itself into the genome of a creature like a shark, carrying with it the code for an enzyme that cuts and pastes DNA. By a fantastic bit of chance, this wayward little gene stuck in an ancestor of modern vertebrates, bestowing on its descendants the potential for a whole new immune repertoire, which has passed down through the eons to our line.

I delight in this idea, that our healing, remarkable, sophisticated immunity may be a kind of memory, offered up by — or perhaps stolen from — the "enemy."

14

MY SALMONELLA

WHEN MY SISTER BECKIE was nineteen, she nearly succumbed to another breed of enemy, a severe case of double pneumonia. I went to visit her in the hospital one April morning and found her lying in a dark room on a respirator, limp from sedatives. Someone had washed her hair the day before, Easter Sunday, and had tucked a pink flower behind her ear. I marveled that it hadn't been crushed. Her arms were stiff, her hands drawn into tight fists close to her chest, her mouth pinched in a tight O. I rubbed her face, stroked her hair, but got no hint of a response. The doctor told me the infection was so widespread in the tissue of her lungs that it had destroyed all but a quarter of one lung.

After six weeks of critical illness, Beckie recovered, surprising us all. That she could battle back from the edge in this way, go on breathing with so little lung, spoke powerfully of the body's resilience and tenacity. For months, though, I wondered why she had become so sick. I cast about for scapegoats — improper diet, negligent caretakers, grief at the death of my mother a few months before — just as my mother herself had searched nineteen years earlier for possible causes of her daughter's microcephaly. But I couldn't avoid my own feeling of guilt at not having attended more closely to Beckie (so that I might have spotted early signs of the illness and somehow staved it off), as my mother had blamed herself, without good reason, for her daughter's disability, believing it was the dreadful consequence of something she had done.

Not long ago scientists reported that a common virus infecting pregnant women can cause mental retardation in the fetus. I seized on this as an explanation of Beckie's microcephaly. Cytomegalovirus, an agent of opportunistic infection, sneaks up when the immune system is weakened and the body is least able to defend itself; it can bring on pneumonia, hepatitis, colitis, encephalitis, and, in a fetus, microcephaly and severe brain damage.

A giant among viruses, cytomegalovirus is named for the way it makes a cell swell and gives its nucleus the look of an owl's eye. A virus, as one scientist said, is just a piece of bad news wrapped in a protein coat. It specializes in molecular thievery. With few genes and proteins of its own, it begs, borrows, and steals from a host cell, with cold efficiency, what it needs to carry on its life cycle and spread to other cells.

Cytomegalovirus belongs to the herpes family and commits a particularly guileful crime, preying on weakness and misleading the immune system by sleight of hand. When a virus reproduces inside a cell, the cell normally displays pieces of the intruder along with its own MHC self-markers. The body's immune cells recognize the viral parts and kill the infected cells. But cytomegalovirus can stop an infected cell from making the markers, effectively gagging it. For those immune cells designed to attack body cells without markers, the virus uses a different ruse. It fashions a fake version of the body's self protein and raises it as a false self-flag, hopelessly fooling the immune system. A mother infected with cytomegalovirus during early pregnancy feels few symptoms. But if the infection spreads to the infant, the virus replicates in the tubules of the kidney and can infect the central nervous system, causing abnormal brain growth, destructive lesions, and severe mental retardation.

I have no evidence that cytomegalovirus caused Beckie's microcephaly. There are no eyewitness accounts, no urine samples or blood smears. It's a long time since the crime. Still, I'm anxious to assign blame, to absolve my mother of her unreasonable guilt, to whisper into my sister's uncomprehending ear, "See, it was something Other, something insidious and unstoppable that made you its victim."

Reading the news, one finds reason to blame microbes for all sorts of ills. Lurking on food, hands, lunch counters, on the knob of the door and the breath of that child sitting so close to mine, we are told, are germs manifold, shifting, and invisible, like ghosts — *Escherichia coli, Salmonella, Staphylococcus* — wishing us ill, it seems, now more than ever. In the spotlight are some old microbial terrors still circling the globe, the perpetrators of meningitis and cholera (which just completed its seventh recorded circumnavigation), along with some grotesque new ones, like Marburg and Ebola, which ravage their victims with grisly symptoms, as well as a few that cause ills we had long blamed on our own bad habits or faulty physiology — ulcer, kidney stones, chronic gastritis, even hardening of the arteries, cancer, and mood disorders. Some microbes, such as the tubercle bacillus, which developed countries thought they had licked, have reappeared with a vengeance in epidemics that sweep through schools and hospitals.

In early use, the word "germ" meant vaguely the seed of a disease. Today, the word signifies the full-blown enemy of our clean, healthy, independent selves. Germs these days are portrayed as the ultimate Others, tiny and potent, literally dying to get inside us and do damage. Some microbes seem diabolically well adapted to attack through all the chinks in our armor, to home in on our cells, latch on to them, and manipulate them in cunning and subtle ways. Take the news about *Salmonella typhimurium,* the agent of typhoid fever, a miniature Iago. The bacterium breathes deception in the ears of our phagocytes, those valiant first-line defenders, convincing them to take their own lives. Or the human papillomavirus, which exploits a weakness in the p53 protein, disrupting the controls on cell growth. Or the whooping cough bacterium, *Bordetella pertussis,* which urges cells in the body's respiratory tract to make a potent toxin that kills neighboring cells charged with the job of sweeping away mucus. The result is the convulsive, spasmodic "whooping" cough that spreads the bacteria to other victims.

From this vantage point, microbes look like things specifically designed to prey on us, to target and destroy our peaceful, passive, innocent cells. But it is not so. The great bulk of bacteria have little interest in humans, set as they are on consuming the leavings of nature, breaking down hard rock and making soil habitable for other organisms, recycling the elements of life. It is the bacterium *Streptomycetes* I have to thank for the sweet, earthy odor of forest I associate with home. The protists — amoebas, dinoflagellates, protozoa — best known for their role in disease, are prime producers of oxygen and food for fish. Even the potentially pathogenic microbes that live in our bodies usually cause no trouble. Most often the relationship is one of peace, shaped by the long push and pull of coevolution aimed at securing the survival of both microbe and host. Half of all pregnant women are infected with cytomegalovirus. *E. coli, Salmonella, Staphylococcus,* hundreds of potentially pathogenic bacteria, and dozens of viruses reside within us as harmless denizens. Only rarely do we suffer because of their presence.

Most of us are infected; few of us are diseased. Microbes generally have the good sense to stay on the surface, in the skin or the lining of organs, where they go unnoticed; those that venture deeper are sometimes invited in by our own cells. Disease is not always a one-sided attack. It may be more like a saraband, an elaborately choreographed dance between the microbes and our cells.

Consider the spread of *Shigella,* a bacterium that causes severe diarrhea. When infected by *Shigella,* host cells give a little gift: they donate their actin proteins to make cometlike tails for their pathogenic guests. Julie Theriot, a biologist at Stanford University, makes videotapes of these microbes as they interact with host cells in real time. She has found that the bacteria use a single gene to coopt a cell's actin and transform it into a bacterial propeller. The bacteria use their new appendages to propel themselves around a cell and out of it to spread to other cells. When Theriot and her colleagues inserted the gene into harmless *E. coli* cells, the bacteria gained the ability to borrow actin tails from their host cells. Whether a *Shigella* infection peters out or turns into raging illness often depends on this exchange. Without their comet tails, the infectious organisms would be stopped cold.

What we think of as disease — tissue destruction — is often the fault of our own bodies. It's not the invading organism that does harm, but our immune cells, which get stirred up, like Othello, by a false reading of messages. The body's occasional overreaction to *Helicobacter pylori,* a normal resident of our stomach lining, may be just such a case. Immune cells pour into the area, piling up, dying, falling apart, spilling their microbe-killing chemicals into the stomach tissue, inflaming the lining, and causing acute gastritis. Food poisoning, another common malady, often comes about when T cells overreact to the presence in the body of toxins left lying about by *Staphylococcus aureus.* The bug is no longer there, just the raging rumor of its presence, but that arouses T cells to multiply wildly and make nasty chemicals that agitate the gut, leading to vomiting and diarrhea. Riled-up T cells also cause the trouble in toxic shock syndrome by making chemicals that open pores in blood vessels, leading to the syndrome's deadly drop in blood pressure.

Our cells are mixing with microbes all the time; only rarely does the encounter move beyond pleasantries to hostile exchange. Most often, the outcome of such meetings is not disease but friendly barter, whereby microbes find shelter in bigger bodies and, in return, offer some service to their hosts.

A shining example from the ocean world is the swap between the bacterium *Vibrio fischeri* and its host, *Euprymna scolopes*, a two-inch squid spotted like a leopard, hunter of night waters above the shallow sand flats of the Hawaiian archipelago. The squid offers its *V. fischeri* food and lodging in a pair of glandlike light organs inside its mantle cavity, selecting that species over all others in the sea. In return, the bacterium gives the squid a unique form of protection against predators attacking from below: an eerie glow, with which the squid camouflages itself by matching the light of moon and stars.

It's a winning bit of fellowship by any measure. When a baby squid hatches from an egg, it is germ-free and bears only a rudimentary light organ. The cells of this organ sweep up *V. fischeri* from the surrounding sea. When the microbes have reached sufficient density, they start to make a light-producing enzyme, luciferase. The squid's light organ then metamorphoses into a full-fledged bioluminescent night-light. So quickly do the bacteria multiply and settle that within hours the young squid can make what light it needs to protect itself from predators. The glow erases the shadow that would normally fall when the moon's rays strike the squid, and in that way camouflages the squid from such bottom-dwelling hunters as lizardfish and eels. The squid can raise and lower the level of light to match the intensity of moonlight by limiting the amount of oxygen that reaches the bacteria. Without the presence of *V. fischeri*, there would be no bioluminescent organ, and the squid would become an eel's easy meal.

I've always had a passion for this phenomenon: two separate creatures joining up, pooling resources, conspiring to make things that neither could have made alone. Some of nature's wildest creativity is constructed from these liaisons.

More than two thousand years ago Herodotus told the story of the plover on the Nile that ate leeches from the mouth of the crocodile. Pliny liked that story and wrote of other partnerships — between peacocks and pigeons, blackbirds and turtle doves, "the crow and the little heron in a joint enmity against the fox kind, and the goshawk and kite against the buzzard." And my favorite: the far-fetched alliance of whale and sea-mouse. "Because the whale's eyes are over-burdened with the excessively heavy weight of its brow, the sea-mouse swims in front of it and points out the shallows dangerous to its bulky size, so acting as a substitute for eyes."

The word "symbiosis," the living together of unlike organisms, was coined in 1878 by Anton De Bary, a German mycologist, for the special case of lichen, a kind of double creature made of a solid partnership between algae and fungi. The algae harvest the sun's energy through photosynthesis and direct the fungi to provide them with a sturdy home. The fungi benefit by the food they obtain from the algae. In this way, a humble growth ekes out life in the toughest conditions — on baking dead rocks in the desert, in the wind and cold high above the timberline, well past the last mosses and alpine flowers, where it grows in delicate rings and whorls etched onto bare granite. It was a similar fusion between leguminous plants and the bacteria that fix nitrogen that brought our green world into being. Without this odd little contract, it's doubtful whether our tribe would be here at all.

We are just learning to see such liaisons; suddenly they're everywhere, a common theme in life. Consider the termite and its resident protozoa, which help the insect digest its woody meals; the tiny violet-spotted Pederson shrimp and the fish it cleans; the yucca and its pollinating moth, a beautiful case of mutualism once remarked upon by Darwin — neither insect nor plant can reproduce without the other's help.

Common is the urge to clasp one species to another, the fig to its wasp, the orchid to its bee. The hairstreaks and blues skimming the weeds and thistles of my yard belong to a butterfly family firmly wedded to ants. It's actually the caterpillar that's the myrmecophile, seducing the ants with acoustic calls and chemical lures that imitate the ants' own siren songs. The caterpillar keeps the ants at hand with a steady supply of the nectar it exudes from a storage organ at its rear. The return favor? The ants shield the caterpillar from its main predator — the wasp, which has the nasty habit of killing with a single sting, cutting up the body, and feeding the meat to its grubs. When threatened by a wasp, the caterpillar releases pheromones from its tentacles, signaling the ants to attack.

Scientists lately discovered an interesting triple alliance of ant, fungus, and bacterium, three members of three separate kingdoms joining forces. Since the nineteenth century it has been known that a kind of leaf-cutting attine ant of South America has a perfect little partnership with a parasol mushroom from the fungal tribe Leucocoprini, which the ant weeds, manures, and propagates to ensure a reliable source of food. But in 1999 biologists found that the coalition includes a third critical member, a bacterium from the genus *Streptomyces*, which makes antibiotics that protect the ant's garden from infection by a common parasite. Could any troika be neater?

So vital are some symbionts that hosts take elaborate measures to pass them between the sexes and from parent to offspring. The common roach contains whole organs made of nothing but bacteria, which are passed down from generation to generation like a beloved Windsor chair. So, too, do aphids transfer their symbiotic *Buchnera* bacteria. Now there's a steadfast relationship. To house the bacteria, each aphid creates a tidy little cellular pocket, the only place where the bacteria can live. For their part, the bacteria make essential amino acids that supplement the aphid's diet of plant sap. Over their long relationship, the bacteria have obligingly evolved extra copies of genes that help make the amino acids. The two creatures will not consent to separation. If prevented by force from mutual association, each will surrender its being.

We humans, too, would languish without our cross-kingdom coalitions. We acquire our biota as does an island, its rock and sand destitute at first, then gradually populated with creeping green. Consider the colonization of my daughter's insides. While in the womb, her body was a pristine wilderness, essentially germ-free. But as soon as she began her short dark passage, she was no longer alone. Riding on the walls of my birth canal, on the sheets beneath my bare legs, on my hospital gown, in the air, nesting on the tips of my nipples, tucked in the creases of my husband's hands and lips, was an ark of tiny organisms, which jumped, catlike, into the cozy habitat of her body. (A baby born by cesarean section tends to acquire slightly different microflora, at least initially — those floating about in a hospital's birthing suite.)

Over the next few days, successive invasions of microbes quickly colonized my daughter, taking up residence in the crevices of her pudgy skin, in the toothless cavern of her mouth, in the tiny tunnels of her gut and vagina, a single-celled tribe with names that make their own odd little poem:

Clostridia, Fusobacteria, Propionibacteria
Coprococcus, Ruminococcus
Peptococcus, Streptococcus
Peptostreptococcus
Bifidobacteria:
B. infantis
B. adolescentis
B. breve
B. longum

Some of these creatures soon disappeared, but many pioneers took hold and multiplied, until, a week later, they formed a community that would grow to be as rich and complex as any found in outdoor nature: four hundred species in the gut, six hundred in the mouth — all ensconced in special niches, inside the cheeks, on the top of the palate, on the back of the tongue, throughout the gut from esophagus to anus. Among them were such exotic varieties as a microbe that normally thrives in lake sediments and one that causes cat scratch fever. *Streptococcus* would come to love especially the plaque on the tiny new teeth that would erupt in three months from my daughter's gum, as would some cousins of *Synergistes jonesii*, a bacterium that also lives in the guts of African sheep, where it serves its host by detoxifying poisons in plants. Its cousins may do the same for us.

These days, the number of individual microbes colonizing my daughter's alimentary tract easily exceeds the number of cells in her whole body. The personal ecology of her skin bacteria, which make their living by fermenting her sweat, helps to determine her particular odor. Bacteria make up 10 percent or more of her body weight. Without them, she would fail.

I once took a course of broad-spectrum antibiotics to fight a bout of bacterial infection after an attack of flu. The drugs killed off not just the guilty bugs but nearly all other resident bacteria. I suffered severe nausea and diarrhea until my normal bugs returned. Researchers have known for years that friendly microbes help digest food, make K and B-complex vitamins, shape the immune system, and keep diseases at bay by competing with — and thus limiting the growth of — pathogenic bacteria. In the Second World War, scientists fed a group of volunteers a diet of nothing but polished rice. The group should have suffered from the vitamin deficiency disease beriberi. But the volunteers all remained healthy for weeks — until they were given a brief course of antibiotics, which killed off much of their friendly flora. Almost immediately, many in the group succumbed to beriberi and other deficiency diseases. Lately, scientists have found that the harmless bacteria living on our tongues have another skill: they make nitrites out of nitrates, providing the key ingredient for a potent chemical made in our stomachs, nitric oxide, which fends off the harmful bacteria we may ingest with food.

The overuse of antibiotics and the hundreds of antibacterial products now flooding the market — soaps, toothpastes, lotions — can wreak havoc with the natural balance of bacteria in the body, at once killing harmless, susceptible bacteria and promoting the growth of so-called resistant microbes. This has led to the emergence of strains of disease-causing bacteria that were once easily stopped with antibiotics but now seem untouchable.

Resistance works something like this. Not all bacteria die from exposure to antibacterial poisons: certain members of an exposed population manage to survive. Sometimes they do so with the help of a mutant gene of their own making or one gleaned from other microbes by means of the little rings of DNA known as plasmids, which can hop from one bacterium to another. In this way, bacteria may gain the ability to detoxify an antibiotic — to modify their own molecules so that the compounds can't react with them or to expel the drugs with "efflux" pumps, like those used by our cells to expunge toxins. Then they may multiply and pass their resistance along to other, dangerous pathogens.

In blocking our exposure to microbes, we may also trigger illness or weaken our immune systems. Whether infection becomes disease sometimes hinges on the time of exposure. If a child is exposed early to Epstein-Barr virus, the immune system learns to tolerate the virus, and it poses little risk. But if first acquaintance takes place during the teen years, the body may revolt at its presence and suffer the high fever and fatigue of mononucleosis. Likewise, late exposure to the bug *Helicobacter pylori* may explain the sudden leap in the incidence of ulcers and gastritis.

To develop fully, an immune system needs to be challenged, exposed to germs that spur the production of the lymphocytes known as Th1 cells.

In too clean an environment, free of challenging microbes, the immune system steps up its production of a different variety of lymphocyte, Th2, the cell that creates antibodies to allergens, a possible explanation for rising rates of asthma and allergies among our children.

So much for autonomy. So much for the virtues of "clean" living.

We are permanent hosts of good guests who contribute to the local economy. "Microbes have lived for a long time without animals," says Abigail Salyers, a microbiologist. "At no point have animals been free of microbes." The bacteria that colonize me are as attuned to my body as any fungus to its fern, orchid to its bee. They have a knack for finding good niches in mammals, having been at it for several hundred million years. When our warm-blooded tribe came into being, microbes knew us for good vessels — warm, moist, ever bent on finding adequate supplies of food and water.

Today their diversity within our bodies is stunning. Our resident microbes differ from those of cattle, pigs, and mice. For that matter, those of a Kansan differ from those of a New Guinean aborigine. When an American goes to New Guinea, the gut organisms of the traveler's home region are challenged and displaced by local varieties, and the result is often something like Montezuma's revenge or Delhi belly.

Our bodies carry more kinds of bacteria than any other living organism, some of them permanent residents, some just passing through. The full spectrum of their members, their number and distribution, remains unplumbed. Of the hundreds of species in the mouth, only half have been identified. Because microbes love company and often cannot live apart from it, they're difficult to study. Even with present technology, it's hard to isolate one from the rest, rear it alone, and keep it alive. Many species have never been cultivated outside the body; they refuse to grow in the lonely expanse of a petri dish.

Although it is largely unknown how these creatures interact with their mammalian hosts, scientists have begun to gather some clues. One key is their sophisticated chemical chatter, according to Jeffrey Gordon, a microbiologist at the Washington University School of Medicine. The lining of the human gut hosts a highly complex society of microorganisms that is constantly talking, secreting signaling molecules to communicate. It's an old habit. During the billions of years that bacteria had the world largely to themselves, they were engaged in just such social liaisons, signaling to one another, working out message systems within and between their communities.

In the mammalian gut, the chemical cross-talk is cacophonous, like the prattle and gab at a raucous cocktail party. To eavesdrop on a single conversation, Gordon uses strains of germ-free mice into which he introduces one microbe at a time. What he finds is that the gut bacteria of mammals are constantly talking not only among themselves but directly to the cells of the body, instructing them to develop and grow, just as *Vibrio fischeri* does with the light organ of its host squid. Every week or two, the body sheds the cells in the lining of the gut to protect itself from natural toxins. New cells are born at the bottom of the crypts of Lieberkühn, a creepy name for tube-shaped pockets in the lining of our small intestines. As the cells mature, they move up from the crypts toward the top of the tiny villi, where they will ultimately help to absorb and digest food. On the trip up, Gordon explains, bacteria supply the maturing cells with chemical signals that help them take shape and perform their various services.

And that's not all. In molding and modifying our cells, the bacteria create favorable niches for themselves and for certain other pioneers. Multiple species of microbes swap chemical messages to encourage the settlement of only select varieties and to oversee the balance between populations of different species, so that a tight, stable net of relations is formed, with an ecology beautifully suited to local conditions. No pathogens need apply.

How strange and wonderful to think that such symbiotic relationships may underlie the normal development and good health of many, if not all, animals. The word "germ" is of doubtful etymology, the dictionary tells me, but perhaps came from the Latin *germen,* offshoot, sprout, or fetus, giving rise to this definition: "The portion of an organic being which is capable of development into the likeness of that from which it sprang: a rudiment of a new organism."

We are finally grasping the etymological hunch here. We know that our brilliant, complex cells are the products of ancient joint ventures between one microbe and another, that the plant cells that feed us came about by like means. We are beginning to talk about germs as life-giving entities, acknowledging that We and They are often in cahoots, colluding, conspiring, merging, and fusing, and that it has been that way for a good long while. For billions of years before we arrived, bacteria blossomed over the globe, cooperated, merged, competed, perfected their molecular wisdom. We are finally admitting that they have taught us to speak *their* tongue, and that our exalted tribe has only lately joined a form of conversation we inherited from them. We're still struggling to grasp the grammar, comprehend the syntax and vocabulary, interpret the word. Perhaps that's why we occasionally misread the signals, get the message wrong, overreact, and cause our own illness.

I don't absolve microbes of their part in terrible disease: the tubercle bacillus or the HIV virus, taking millions of lives a year; *Streptococcus pneumoniae,* destroying the tender tissue in the lobes of the lung; or cytomegalovirus, ravaging the brain cells of a baby. But the seeds that sicken and slay us are the strange, exceptional cases in a generally friendly, indeed indispensable, lot.

The lines between ourselves and our microbes are not solid but slim and dotted. Matched gene to gene, protein to protein, we make connections, cast our lot with creatures from another kingdom. I'm glad to host the billions of microbes that make me one, to think of their numbers as another kind of family, glad to know that the same bacterial chatter that makes the squid invisible against the moonlit sea also feeds, protects, and doctors me.

15
ROOT TALK

> All living things have much in common. . . . We see this even
> in so trifling a circumstance as that the same poison often
> similarly affects plants and animals; or that the poison se-
> creted by the gall-fly produces monstrous growths on the
> wild rose or oak tree.
>
> — Charles Darwin

MOST OF MY LIFE I have been subject to classical migraine headaches, a
syndrome that runs in my family. For years I carried around small brown
pills, one of which I would take at the first sign of an attack. It would begin
as a firefly of light, a pulsing white dot or hole in my field of vision, noted as
the disappearance of a word or two on the page, or a tiny pocket of loss in a
familiar face. Then the hole would grow in size and luminosity, expanding
into a zigzag of brilliant white light — a so-called fortification spectrum —
which advanced at its jagged, prismatic edges until half my field of sight
disappeared in a boiling void. After fifteen minutes of blindness, a tingling
would begin in the little finger of my right hand and creep up from my
palm like a swarm of ants, until I lost all feeling in my right arm and shoulder. When the numbness reached my head, I would lose my words in
strokelike aphasia and grow confused. I couldn't walk a straight line or tell
you my phone number. Unless I took one of those little brown pills, the
aura would give way to chills, sweating, roiling nausea, and a unilateral
headache of blinding severity that might last up to forty-eight hours.

Hildegard of Bingen had migraines, accompanied by visions of great
beauty and rapturous intensity, "clouds of living light" and fortification
spectra, which the twelfth-century nun interpreted as the *aedificium* of the
City of God. Though the word "migraine" has come to be synonymous
with any severe headache, the classical migraine describes a particular con-

stellation of symptoms, from the shimmering visual disturbance to the blinding one-sided headache so great that to sufferers the only solution seems surgical removal of half the prefrontal lobe. Caesar, Darwin, Freud, Kant, Lewis Carroll, and Thomas Jefferson had migraines. So did my father.

I had my first attack when I was twelve; it came on one spring day in a peaceful patch of woods sweet with flickering sunlight and the strong scent of pine, where I sat eating lunch and contemplating my parents' recent decision to get divorced. Almost anything can trigger a migraine in those who are susceptible: fatigue, allergy, strenuous exercise, humid weather, smog, blaring noise, certain vibratory images such as striped wallpaper or falling snow, menstruation, monosodium glutamate, strong cheese, or emotional upheaval at, say, the news of parental schism.

One inherits a predisposition to the disorder, although fifty years of investigation have not illuminated just what that predisposition is, whether it is a single gene or the vaguer notion of a "constitutional type." In 1959, Dr. W. C. Alvarez wrote down what he saw as the characteristic traits of migrainous women:

> . . . a small trim body with firm breasts. Usually these women dress well and move quickly. Ninety-five percent had a quick eager mind and much social attractiveness. . . . Some 28 percent were red-headed, and many had luxurious hair. . . . These women age well.

Some still believe there is a migrainous type of woman, though not nearly as attractive as Alvarez's model; she is most often described as ambitious, perfectionist, cautious, obsessed with order, emotionally constipated.

A migraine attack starts in the cerebral cortex as a sort of slow seizure and spreads like a ripple in a pond, kindling the aura, the visual disturbances, then migrating to the nerves of the blood vessels, glands, and viscera, causing the nausea, sweats, the flush of blood into cerebral arteries, and the piercing vascular headache. Its chemistry is complicated, with changes in nearly all the neurotransmitters that are present in the brain: adrenaline, acetylcholine, histamine, and especially serotonin. Oliver Sacks, who has written so beautifully on the eerie metaphors to be found in the afflicted human mind, believes that migraine may have originated during the course of evolution as a response to such physical threats as exhaustion or illness, injury or pain, not unlike the passive protective reaction of other animals: the curling-up of a hedgehog or the sham death of an opossum. I find it strangely comforting that there may be some buried purpose to this "contumacious and rebellious Disease," as one seventeenth-century physician described it, "deaf to the charms of every Medicine."

These days there are drugs that can stave off the headache. The latest varieties do so by tinkering with the brain's level of serotonin. But for years the main remedy was in the little brown pills I carried, known as cafergot. Taken at the first sign of an attack, they constrict the swelling blood vessels surrounding the brain, a miracle wrought by their blend of active ingredients, caffeine and ergotamine.

For migraine, essence of coffee bean, blended with ergot, a poison made by the small purple-black fungus *Claviceps purpurea*. For the more common tension headache, a derivative of salicylic acid, found in meadowsweet, madder, wheat, and willow tree. For deep, difficult pain, the solidified juices of the opium poppy. For malaria, quinine from cinchona bark. For heart failure, the juices of *Digitalis purpurea*, the purple foxglove plant. Colchicine from meadow saffron for gout. Reserpine from the snakeroot for high blood pressure. The leaves of the coca plant to relieve hunger pains and provide stamina.

The world is awash in potent chemicals. To fight foragers and pathogens, plants and fungi make alkaloids, glycosides, tannin, cyanide, aflatoxin, resins, gums. It's lucky for us that they do. In the early 1900s, 80 percent of all medicines were obtained from roots, barks, and leaves. Even today, a quarter of prescription drugs contain ingredients that are derived from, or modeled on, substances made by plant or fungus.

That plant ingredients have specific and powerful effects on the human body has long been common, if largely rustic, knowledge. Sixty thousand years ago, in the Shanidar cave in the Zagros Mountains of northern Iraq, an adult male *Homo sapiens neanderthalensis* was buried ritually in the fetal position, knees to chin, together with his ax and an arrangement of flowers identified by pollen analysis as hollyhock, yarrow, St. Barnaby's thistle, ragwort, grape hyacinth, and woody horsetail. Several of these plants have medicinal properties and are still used by traditional healers in the region: grape hyacinth as a diuretic, parts of the hollyhock for relief from spasms and toothaches, the leaves of the woody horsetail shrub,

Ephedra alata, as a remedy for asthma. The *Ephedra* shrub contains ephed-rine, a compound that narrows blood vessels in the membranes lining the nose; in Western medicine, it is commonly used as a decongestant.

Seven thousand years ago, the Chinese had sophisticated botanical pharmacopoeias, as had the Babylonians and the Assyrians. The *Pen Ts'ao,* written by the herbalist Shen Nung in 2800 B.C., lists 366 plant drugs, among them ephedrine. Ancient Hindu seers urged the use of *Rauwolfia serpentina,* the snakeroot, to treat nervous disorders and mental illnesses, as well as dysentery, cholera, and fever. In the fourth century before Christ, long before the three wise men offered myrrh to the newborn Jesus, Hip-pocrates prescribed the secretions of the thorny, flowering shrub for mouth sores. Hippocrates also held in high regard the salicylate plants for relieving pain and fever, and he recommended topical application of willow leaves as an antiseptic. The Greek surgeon Dioscorides, whose *De Materia Medica* from the first century dealt with some six hundred medicinal plants, pre-scribed the foxglove plant as a cure for everything from colds to dropsy. A century later, willow juice inspired fierce veneration in Galen, physician to Marcus Aurelius and author of a monumental thirty-volume pharmaco-poeia of remedies — most of them vegetal. Native Americans also made good use of willow bark, decocting drafts to ease rheumatism and fever, and relied on blue cohosh as a cure for menopausal symptoms.

The ancients suspected that humans were not alone in knowing the power of rhubarb and fern. "Animals, too, have discovered plants," wrote Pliny, one of Galen's contemporaries:

> Before giving birth [female deer] use a certain plant called hartwort as a purge, so having an easier delivery. . . . Celandine was shown to be very healthy for the sight by swallows using it as a medicine for their chicks' sore eyes. The tortoise eats *cunila,* called ox-grass [perhaps pen-nyroyal], to restore its strength against the effect of snake-bites; the weasel cures itself with rue when it has had a fight with mice. . . . The stork drugs itself with marjoram in sickness, and goats use ivy.

Montaigne noted that a goat wounded by an arrow will pick dittany out of a million herbs for its cure, and that a tortoise, when it has eaten viper meat, looks for marjoram to purge itself.

Three hundred years later, the Harvard anthropologist Richard

Wrangham noticed that wild chimpanzees in Gombe National Park in Tanzania showed an unusual interest in the leaves of *Aspilia* plants, relatives of the sunflower. The chimps chose only young leaves, Wrangham noted, rolled them around on the tongue for several seconds, and then swallowed them whole. The meal clearly was neither gastronomically pleasing nor nutritional: the chimps ate with a grimace, and the leaves turned up whole in their feces. *Aspilia* plants contain red oils with potent compounds that kill worms, fungi, and viruses. Healers in Nigeria, Cameroon, and Ghana use the plant to clean sores and to relieve coughs and stomach troubles. The Shambala people of eastern Tanzania use it to treat diseases of the nerves.

Many animals eat plants with medicinal properties. Elephants, bears, wild dogs, rhinoceroses, colobus monkeys, pigs, civets, jackals, and gerbils all consume medicinal herbs and grasses, possibly as painkillers or antidotes for poison or to eliminate intestinal parasites. Skeptics wonder whether these are truly instances of self-medication. It is hard to prove that an animal is using a specific plant drug for a specific purpose. But it's an accepted fact that rats eat clay to induce vomiting after they've consumed poison. Starlings searching for nesting materials pick yarrow, wild carrot, fleabane; and other fresh vegetation loaded with chemicals that kill mites and other avian parasites. Formosan termites fumigate their subterranean nests with naphthalene, the powerful chemical with which Jean-Henri Fabre tried to mask the sex attractant of the peacock moth. The termites use the chemical as a defense against ants, nematodes, and pathogenic microorganisms. It's also well known that monarch butterflies feed on the poisonous *Asclepias,* the common milkweed, as a form of protection. The butterflies incorporate the plant's cardiac glycosides into their tissues, and the poison provokes vomiting in blue jays and other predators, which eventually learn to abstain from the butterflies.

If, as Pliny suggested, the Romans learned herbal remedies from the swallows, I wonder whether wild animals may not have guided a Neanderthal healer both in choosing his potion and in avoiding his poison.

Galen held that herbs work by correcting an imbalance among the four bodily humors. Not so, said Paracelsus, father of medicinal chemistry. Like the ancient Chinese, Paracelsus believed that plants are far more specific in their action. For every human ailment there is a particular plant remedy, he claimed, the clue to its efficacy lying in the plant's physical traits — shape, color, taste. Like would affect like: heart-shaped leaves for diseases of the heart, yellow flowers for jaundice, quaking aspen for palsy. Nearer the mark was the seventeenth-century poet George Herbert, who wrote that "herbs gladly cure our flesh because they find their acquaintance there" — words that are more than just mystically true.

Among the hundreds of plants cultivated or gathered by the Neolithic lake dwellers of central Europe was *Papaver somniferum,* the ravishing red poppy flower, with its broad papery petals and its stamens that quiver like a sea anemone. Six thousand years ago, the Sumerians of the Tigris-Euphrates basin, near present-day Iraq, called the juices of the flower "lucky" or "happy."

When my mother was dying of cancer, *Papaver somniferum* was a godsend. My mother was a stoic woman, able to bear the pain of an ear infection, a crushed vertebral disk, an oven burn, to see that we got to school on time or had the bedtime story we needed. But in the last stages of her cancer, she protested the severity of her pain, an awful, deep, burning that cried out for frequent injections of opiates. I tried to deliver the drug myself, practicing on oranges, plunging the needle through the thick, fragrant peel. But I couldn't make the leap to flesh and so watched jealously as the nurse delivered my mother narcotic peace.

Opium's active ingredient, morphine, was isolated in 1803 by a German pharmacist, who named the new substance after the Greek god of sleep and dreams. Not only did the drug ease pain — morphine has the power to raise one's pain threshold by about 70 percent — but it blocked the passage of stimuli conveying tension and anxiety. Here was a comforting remedy that could quiet infants, induce sleep, suppress coughs, elevate mood, ease the anxious, relieve pain. The isolation of morphine, coupled with the invention of the hypodermic syringe in the mid-nineteenth century, meant that the drug could be delivered in pure form and in high doses, a combination that quickly led to problems of addiction.

Why would the cells of our nervous system react to the juices of the poppy or the molecular messages in the coffee bean or in the bark of a willow tree? Why would our hearts jump in response to a stimulant made by the foxglove plant?

The Mysteries of Opium Revealed, a treatise on the drug published in 1700, precociously suggested that opiates act on specific areas of the brain involved in the sense of euphoria. Two and a half centuries later scientists hit on the idea that the drugs might somehow fit into specific receptors on the membranes of nerve cells in the brain. Just such receptor "homes" were discovered in the 1970s, proteins tucked into nerve cell membranes that fit neatly with opiates, like lock and key. These receptors are found not only in the brain, but throughout the body — in the spinal cord, in the uterus, even on the surface of our white blood cells — which explains why morphine exerts such a sweep of effects in our bodies, including the tendency to suppress immunity.

We're not unique in possessing receptors that fit neatly with plant narcotics. Apes get high on opiates. Monkeys, rats, and mice respond to cocaine by bobbing their heads and becoming hyperactive. Jay Hirsh and Colleen McClung, biologists at the University of Virginia, found that cocaine can affect a much smaller creature in much the same way it affects larger organisms, including people. In 1997 the two scientists exposed fruit flies to "crack" cocaine vapors and videotaped the flies' behavior. The films are disquieting. At low doses, the flies groomed themselves continuously. At higher doses, they walked backward, sideways, and twirled in slow circles — highly unusual behavior for flies but similar to that observed in coked-up mammals. At the highest doses, some flies developed tremors and paralysis; others died. Many drugs, especially those linked with abuse, seem to act on ancient, conserved neural mechanisms — common brain circuits for reward. From their studies of fruit flies, Hirsh and McClung are hoping to gain insights into the nature of addiction in humans.

The puzzle of why the cells of the human brain — or, for that matter, those of a fruit fly — might have receptors for the distillations of a plant was solved in 1975. Two Scottish researchers found in the brains of pigs a chemical that acted like morphine, the first of many so-called endogenous opioids to be discovered; they were later named endorphins, for "morphine within." Opiates stop pain the same way our endorphins do, by locking on to opiate receptors on the surface of nerve cells, preventing the pain message from traveling to the brain. The difference between my stoicism and my mother's may lie in the number and working of our opiate receptors.

This discovery suggested a new idea: many plant molecules affect us because they mimic our natural chemicals and thereby fit neatly with our receptors. Blue cohosh and other plants make "phytoestrogens," substances that lock on to our receptors for the hormone estrogen. Myrrh's active ingredients bind to the brain's opiate receptors. The chemical THC in marijuana docks with our receptors for the brain chemical anandamide. The brains of mammals contain high numbers of these so-called cannabinoid receptors, which suggests that our internal cannabinoid-like substances and their receptors may be important in the brain — although their role remains unknown.

At first, the likenesses between plant compounds and endogenous chemicals were chalked up to coincidence. Later, it was believed that plants had evolved such mimicking neuroactive substances purely for defense against foragers. Then, in 1998, a team of Chinese and American scientists proposed another explanation. In a kind of flowering plant — a wild mustard called *Arabidopsis thaliana*, which is the darling of laboratories, the plant equivalent of the fruit fly or the mouse — the team found two genes that made receptors for glutamate, much like the glutamate receptors involved in the signaling between nerve cells in the human brain. (In the brain, glutamate acts as a chemical messenger, playing a part in acquiring and storing memories; its failure is implicated in Alzheimer's disease.) Finding the receptors was a stunning surprise; scientists had thought these genes only existed in animals with nervous systems. The glutamate and its receptors, it turns out, play a vital role in the plant's inner life, helping it to detect and process light signals.

It may be, then, that our cells respond to the molecules made by meadowsweet and marijuana not only because herb and herbivore co-evolved in a kind of evolutionary arms race, but because we share primitive mechanisms of signaling and receiving that existed before plants and animals diverged.

This language likely arose when life began. To signal one another, the earliest organisms invented a suite of messenger molecules — chemicals with the power to inflame, dilute, cancel, inspire, calm, alter, or deflect one another's effects — enabling one cell to communicate with another and affect its behavior. As organisms evolved, the signaling molecules became more elaborate through gene duplication and divergence, allowing multicellular organisms to send specialized signals between widely separate parts of the body and out into the world. Among them were neurotransmitters and hormones, which, released into the bloodstream of animals or the sap of a plant, carry signals to distant target cells. These ancient biochemical messengers have remained in place for hundreds of millions of years, enabling plants and animals to instruct their cells in proper behavior, to grow and defend themselves, find food, avoid danger, convey alarm, mark territory, locate home, distinguish friend from foe, announce sexual readiness, and, yes, defend themselves. It may be to the conservation of this ancient chemical vocabulary that we owe our response to caffeine, ergotamine, and aspirin, and our ability to gain relief from the pain of cancer through a steady dose of opiates.

Wonderful, healing substances have come from the natural poisons, toxins, and venoms made by other organisms — chemicals auditioned by life and often more effective than those cooked up in a laboratory. A compound from the skin of the frog *Epipedobates tricolor* has given us the potent painkiller epibatidine, which works by a mechanism completely different from that of morphine and lacks the opiate's drawbacks. From the complex cocktail of one kind of cone snail, *Conus magus*, scientists isolated a compound that provides relief from the pain of cancer, nerve damage, or amputation by interrupting pain signals as they travel through the spinal cord to the brain. The drug paclitaxel is made from the "spindle poison" taxol, found in the bark of the Pacific yew tree. Taxol kills cancer cells by making the microtubules rigid so that cells die as they divide. Vincristine and vinblastine from the dainty Madagascar periwinkle also block cell division, checking the growth of tumor cells in childhood leukemia and quadrupling an afflicted child's chance of survival. Betulinic acid, drawn from birch bark, can slow the growth of human melanoma by inducing apoptosis. Michellamine B, isolated from the leaves of the Cameroon liana, is a promising agent against the AIDS virus.

There are other antidotes out there, in the black waters of the deep sea, in the soil of the temperate forest, and in the rainforest, where the biggest snakes in the world have their kingdom in a dark green tangle of hundreds of thousands of different plants. Only a tiny percentage of bacteria, fungi, and plants have been named, much less sampled for their compounds with the ability to heal — adequate reason by itself to insist on preserving all the species of life's tapestry, whether to our eyes magnificent creatures or humble slime molds.

"There is no animal in the world exposed to so many injuries as man," wrote Montaigne. "Lice are enough to vacate Sulla's dictatorship; and the heart and life of a great and triumphant emperor is the breakfast of a little worm." We may be at the top of the food chain, removed from the game of predator and prey, but we are still subject to pain, still consumed from the inside out, still plagued by diseases embedded in our genes. So do we fly toward those ingenious cures cooked up by our companion creatures, which speak to our own cells in a common chemical cross-talk. So do we scavenge the rainforests and coral reefs for potent new ones, purify them, pour liquids up and down, one into this, another into that — adder's fork and blind-worm's sting, lizard's leg and howlet's wing — brewing mixtures from sunrise onward that we may dim our pain and stay our sojourn a little longer.

PART IV

PASSAGE

From the Indo-European *pet-*, to spread. Going through, over; a journey, especially one by air or water; passing from one state or stage to another. Obsolete: death.

16

THE LOOP OF TIME

EARLY ONE MARCH MORNING, I was standing in the dark of my yard with my four-year-old in my arms. The ground was hard with the crust of late winter, and the wind was cold, but there were faint notes of spring, sweet pockets of warmth emanating from the soil, and a lone daffodil in bud, which lifted my spirits.

I had come out with my daughter Zoë in the small hours long before sunrise to witness the coming of a comet. Hale-Bopp was journeying out from the Oort Cloud, a dark body in a remote region of the outer solar system. It was first discovered a few years ago beyond the orbit of Jupiter, and now, roaring through space at close to a hundred thousand miles per hour, was nearing its closest approach to the sun.

I stroked Zoë's forehead to urge her awake. I suppose I had in mind the old, passed-down stories of young children raised from their beds in 1910 to witness the visit of Halley's comet in the hope that they might glimpse the comet twice in a lifetime.

No chance of that here; Hale-Bopp would swing out to space for another three thousand years. Still, the comet was a source of great excitement, a monster with an icy core twenty-five miles in diameter, and no mere dazzling astronomical show but a messenger from another time, a frozen relic from the birth of our planetary system. As it neared the sun, astronomers said, the comet was warming and disgorging its molecular secrets, which were encrypted in the tons of ice and dust grains now evaporating and fanning out from its tail: molecules of frozen water, methane, methyl alcohol, methyl cyanide, formaldehyde — chemicals in which scientists hoped to read the beginning of sun, moon, Earth, time.

It seemed well worth getting up in the bottom half of night. But so early was the hour that the mockingbird was still locked in sleep, and the owl that had lately taken possession of the bird box on our maple tree had not yet returned from its night hunt. I was still in a mind and body fog, on the cusp of consciousness, and I searched the starry sky in sleepy confusion, not sure what I was looking for. I had seen a map of the comet's location marked amidst neat, orderly constellations — Perseus, Cassiopeia, Lacerta, Pegasus — but the sky was to me a chaotic glut of stars.

Finally, low in the northeast, I spotted a faint spray of celestial light. I shifted Zoë to my hip, rubbed my eyes with my free hand, and raised my binoculars.

Few things in time or space are more ghostly than comets, icy clumps of stardust rushing at us out of blackness at frightful speed, trailing a fire that doesn't burn, looping close to our planet, then vanishing again into space, to return at an appointed time decades, centuries, millennia later. Comets are reassuring in one sense, reminders that the depths of space hold water, carbon dioxide, nitrogen, elements the same as those found on Earth. But they are also alarming. They have shattered the planet in the past and may do so again. Comets were once feared as harbingers of doom and death — and still are by some. With the appearance of Hale-Bopp, thirty-nine people took their lives, believing the bright icy ball both portent and ticket to paradise.

At the speed it was traveling, I expected to see this comet move, but it just hung there, low on the horizon, a comet at anchor. From the papers I had learned that Hale-Bopp at its nearest point to Earth was still more than a hundred million miles away. The sunlight it reflected, traveling toward us at 186,000 miles a second, would take nearly eleven minutes to reach our eyes. We were seeing the comet not as it is, but as it was. This is a concept I first learned from my father, who taught me while bird watching in the pre-dawn hours that astronomers look only at old, far-traveled light, that time is born of space and owes its existence to the expansion of the universe, to a single elusive moment in some small domain of colossal temperature and nearly infinite density in which all was set in motion.

At the moment I could think little of physics, only of the cold, dark quiet of late winter, this funny unmoving smudge, and the sleepy warmth of my daughter, whose half-moon eyes were hardly taking in those slow photons. My body was tingling with fatigue, remembering an old rhythm I was violating. I'm a lark as these things go, as is my father, but this hour was unnatural even for our line — the hour of the wolf, Ingmar Bergman called it, when warmth retreats and clots thicken, the hour when members of our species are likeliest to die or be born, to have migraines, asthma, joint pains, heart attacks, to mourn mistakes, dread the perils of the un-folding day, feel despair.

As surely as the time of day dictates the mockingbird's waking call, the hour defines my internal state — the waxing and waning of pulse rate, blood pressure, temperature, the size of my airways and patterns of breathing, my populations of white blood cells, levels of iron, sugar, and zinc — and the condition of consciousness, my grasp of astronomy and physics, my sense of well-being. All ebb and flow in a daily rhythm close to, but not the same as, the Earth's rotation (hence the word "circadian," from the Latin *circa*, about, and *dies*, day).

In those dark early hours, my immunity was weak, but my flow of prolactin, the hormone that triggers the making of milk, was peaking in readiness for the morning hunger of Zoë's baby sister. At dawn my pulse rate and blood pressure would rise sharply and the platelets in my blood grow sticky. From its predawn lowpoint, my body temperature would rise two degrees over the next twelve hours. At midday my blood level of hemo-globin might soar, but my mind and body would sag in an hour of letdown, a slump parallel to early morning, when sleep would beckon and spirits flag. By midafternoon my grip would strengthen, body temperature and blood pressure rise; my tolerance of pain would peak, and, a few hours later, my tolerance of alcohol. As the outside world grew dim and illegible, my lymphocytes would multiply and my mind droop and close its petals, like a morning glory at dusk.

This daily up, down, in, out, rhythm, tuned to the cycle of day and night, is about as universal as any bit of biology gets, occurring in every plant and animal, in ermines and emus, in koalas and kings, even in single-celled algae and bacteria. Life arose under the influence of the Earth's rota-tion, a reigning power that has dictated exposure to solar and lunar cycles over the eons. It's little wonder that our biological rhythms swing in syn-chrony with its forces.

As early as the fourth century before Christ, Androsthenes, scribe of Alexander the Great, noted the opening of tree leaves during the day and their closing at night. Most creatures have lives turned either to noon's face or to midnight's. The owl hunts in darkness and sleeps in daylight, mimicking the pattern of its rodent prey. The daffodil lifts its leaves in the day and takes up nutrients from its roots at night. The heliotrope's leaves open in the day and fold at night; its flowers trace the arc of the sun, facing east in the morning and west in the evening. (So taken was Linnaeus with the discovery that the petals of many flowers opened and closed at specific times that he created a floral clock, in which morning glories and lilies told the time.) Even the concentration of codeine in opium poppies varies regularly with the hour.

Early observers assumed that such rhythms in nature were strictly controlled by the pulse of sunlight, that an organism's clock obeyed the dictates of light and dark. But in 1729 the astronomer Jean-Jacques d'Ortous de Mairan disproved this notion in an experiment with a heliotrope plant kept in the dark. The leaves and flowers of the heliotrope stuck faithfully to their daily routine of opening and closing, even when locked in perpetual night.

The free-running clock is widespread among organisms. Darwin noted that earthworms removed from all sources of light remained in their burrows during daytime and emerged at night. In the 1930s, Karl von Frisch found that honeybees set out to forage each day at the same time, even under steady conditions of artificial light, in strict accordance with the motions of a hidden sun. Fruit flies reared generation after generation with no natural light burst from their pupal cases just when they should, precisely as the sun rises. Even the humble potato, snatched from its environment and deprived of light, sticks to its cycle, consuming oxygen in a neat, reliable circadian rhythm. Our bodies, too, know the length of the daily cycle. Enclosed in a dark room with no windows or clocks, we maintain our patterns of sleeping, waking, activity, hunger, fixed to the twenty-four-hour cycle.

There is something ticking inside all of us, generating biological night and biological day at precise intervals, telling plants when to grow, rest, lean toward light, reminding insects when to emerge from their pupae, inciting us to wake with the day and sleep with the night. This something keeps itself wound and does not neglect to keep time even when locked in the dark, yet is so sensitive to light that a single flash 1/2000th of a second long can shift its rhythm. It is so flexible that its "hands" can be bumped forward or backward each day to stay in harmony with the changing day lengths of the seasons, so that in summer biological night is short and in winter it is long — as if we all had caught a planetary rhythm and internalized it so that we mimic celestial oscillations even in the absence of their cues.

In the 1950s researchers looking for an internal timepiece in mammals set about trying to destroy it. Convinced that a body clock must be big and complicated, like the one in my grandparents' hallway, and therefore be seated in a major organ, they surgically removed from rats and hamsters various organs — the pancreas and gonads, the adrenal, thyroid, and pineal glands. They tried disrupting the animals' clocks with anesthetics and alcohol. All to no effect. Finally, after decades of work, they found they could alter a rat's rhythms by damaging its suprachiasmatic nucleus, a tiny knot of about ten thousand cells in the hypothalamus, at the bottom of the brain. A similar "master" clock existed in the pineal gland of birds, amphibians, and reptiles and in the retinas of marine organisms. It was assumed that biological rhythm arose from a complex interconnected system of gears and cogs in the body, all controlled by this master clock.

Then, biologists studying vision in the sea hare stumbled on a curiosity. The sea hare is a marine snail of sorts, with a body shaped like a rabbit's and two prominent tentacles suggestive of ears. The scientists noted that the optic nerve in the eye of the sea hare fired in a regular circadian rhythm. But when the scientists removed the eye, and kept it alive in a solution of essential nutrients, they were shocked to find that the daily rhythmic firing of the nerve persisted.

Intrigued by this phenomenon, Gene Block, a biologist at the University of Virginia, went on a deeper search for the rhythm's source in a relative of the sea hare, *Bulla gouldiana,* the cloudy-bubble snail. The eye of *Bulla* is made of a lens, along with thousands of photoreceptor cells and other neurons. In a delicate operation, Block whittled away at the eye, first shaving off the lens, then all of the photoreceptor cells, until only a hundred or so neurons were left. These he removed cell by cell until a cluster of only six cells remained. The rhythm persisted. Eventually Block carved his way down to a single cell. To his astonishment, that isolated cell continued to beat all by itself, its reliable rhythm perfectly synchronized with the swing of night and day.

An individual cell, it turned out, could undergo daily cycles of activity and rest, just as whole organisms do, without any help from other cells — keeping beat, it was soon discovered, by means of a tiny ingenious design, old, universal, and of vast importance.

The word "time" gets more space in *Webster's Dictionary* than "life," more than "good," "evil," "god," half again as much as "truth," twice as much as "love." And that's not counting the myriad compounds — time-honored, timeworn, time lag, time being, timeless. We are a time-minded tribe, mincing our days into hours, fretting our seconds. In the fourteenth century, King Charles V of France, tired of hearing the church bells of Paris at irregular intervals, ordered a uniform time for the city, making the clocks synchronous with the master clock in the tower of the Palais Royal. Another two hundred years would pass before the Dutch physicist and astronomer Christian Huygens turned the predictability of an oscillating bob into a means of reliable timekeeping, sending us on a quest for absolute precision in measuring the passage of time.

On New Year's Eve in 1998, the world's official timekeepers at the International Earth Rotation Service in Paris added a second to the 86,400 seconds that mark the Earth's rotation. Leap seconds became necessary in 1972, after scientists redefined the standard second as the time it takes for a cesium atom to vibrate 9,192,631,770 times. A half-century earlier, the National Bureau of Standards had unveiled the first clock regulated by the vibration of an atom — specifically, the oscillations of one nitrogen atom in an ammonia molecule, which are counted off electronically to generate the ticks of a clock. At first the new atomic clocks produced one second of error in three hundred years; these days the error rate is one second in six million years.

So accurate is our timekeeping that it no longer agrees with the rotation of the Earth, which bumps and wobbles on its axis like a wavering top, slowing down and speeding up by small amounts, depending in part on friction caused by tides. Now and again, the wizards of time add a second to keep clock time matched to the Earth's erratic spin. If they didn't, ten thousand years from now our clocks would show the sun rising at noon.

We are obsessed with marking time accurately, as if time were something we could measure with a ruler; as if doing so would make it behave as we wish. But no matter how loud the chime, time flows by flux of mood. Certain mystics in moments of illumination are said to experience, in only a few chronological seconds, years of transfigured love. When I'm wandering the seashore, time is a caesura. When I'm in the vice-grip of migraine, it's an epic. The passage of time may dissolve some memories, but it does nothing to dispel the loss of a loved one.

We have learned from the physicists that no matter how accurately we measure it, time is not an absolute quantity. Relativity decrees that a clock in motion runs slower than a clock at rest. At the speed of light, time stands still. What makes time an arrow is entropy, the coming apart of all things, the universal trend toward disorder. And, say the physicists, cosmic time, like personal time, is not without beginning or end but like the sands in an hourglass, in finite supply. Time is running out.

It is news of this sort that sends me in search of a small stable sanctuary, sets me digging about for bits of biology that have gone unchanged through the long beat of the Earth's millennial heart.

When biologists first tried to trace the source of an animal's internal ticking down past the cell to the molecular level, they faced certain difficulties. As earlier researchers had noted, circadian rhythms are difficult to destroy, resilient as they are to physical or chemical disruption. They're neither slowed down nor speeded up by heat, which affects most biochemical reactions inside cells. With an increase in temperature of 20°F., some reactions have been found to go four times as fast. But even with temperature swings of 30°C., body clocks maintain their normal tick. They're equally immune to a brew of disruptive chemical agents, metabolic and respiratory poisons, photosynthesis inhibitors, growth factors, and other insults that undo many biological functions. The trick was finally accomplished with the potent chemical ethylnitrosourea, choice weapon of developmental biologists for mutating genes. Now scientists were able to use genetic screening of mutants to dissect circadian clocks in organisms from the fungus *Neurospora* to fruit flies.

A major breakthrough came in the 1970s. When mutated, a gene called *period*, or *per*, undid the daily rhythms of fruit flies, making the insects emerge from their pupal cases at odd hours. Other genes that affected the fly clock eventually turned up, including one called *Timeless*, or *Tim*.

But knowledge of these fly genes did not help scientists find the timekeeping genes in mammals. That required a serendipitous discovery by Michael Menaker, then a biologist at the University of Oregon. In a standard laboratory shipment of Syrian hamsters, Menaker and his colleagues found an animal with a disrupted circadian rhythm. The hamster had a shortened twenty-hour rhythm of sleeping and waking caused by a genetic mutation they called *tau*, the Greek letter and symbol for "period." When cells from the suprachiasmatic nucleus of *tau* mutants were transplanted into normal hamsters, the recipients adopted the twenty-hour rhythm, highlighting the importance of the organ in mammalian circadian behavior. But the molecular identity of the *tau* mutation remained elusive.

Then, in the early 1990s, Joseph Takahashi, a neurobiologist at Northwestern University, searched for mammalian timekeeping genes by working from behavior back to genes in mice. He planned to look for mice with broken biological clocks and examine their genomes for the broken genes that might have caused the aberration. Takahashi got a shipment of mice whose parents had been served meals laced with mutagens to disrupt the genes of their progeny. He watched the hundreds of offspring run, hoping to find one with a crazy clock. Laboratory mice like to rest in the day and run on exercise wheels at night and will start running at precisely the same time each night. Takahashi got lucky: the daily rhythm of mouse number 25 was long by an hour. Descendants of that mutant mouse, bred to have two copies of the mutant gene, displayed a circadian rhythm about four hours longer than normal. It took Takahashi and his team of ten researchers three years to fish out the gene responsible for the untimely behavior, a giant made of a hundred thousand base pairs, which they named *clock*, for "circadian locomotor output cycles kaput." In the spring of 2000, the team announced that it had also laid hands on the *tau* gene.

We humans have versions of *clock* and *tau* genes, too, similar to those in the mouse and hamster. *Clock*, it turns out, makes a protein that shares a key feature with both the per proteins of fruit flies and the timekeeping proteins of bread mold — a common protein domain known in the Esperanto of the laboratory as PAS. Similar versions of the fly clock genes *time-less, doubletime, cycle, cryptochrome,* and others have turned up in frogs, fish, mice, and humans, suggesting that the genes first appeared in a common ancestor some time before the Cambrian explosion, more than a half billion years ago.

Small variations in our versions of these clock timekeeping genes may spell the difference between larks happily up at dawn and those more owlish in bent, who hit their stride at midnight. Mutations of the genes may also explain the more disruptive body clock problems that run in families, such as advanced sleep phase syndrome, which sets a body three or four hours ahead of the norm, inducing sleepiness at twilight and wakefulness well before dawn.

In 1997, scientists hit on an ingenious way to watch these genes in action. By harnessing fruit fly *per* genes to genes that give fireflies and jellyfish their fluorescent glow, the scientists could see the *per* genes literally flash their activity. To their surprise, they found that the genes were active everywhere — in the flies' heads, thoraxes, abdomens; in disembodied probosci and antennae — independent clocks in individual cells keeping time in the absence of a brain and responding directly to light. Since then, scientists studying the molecules and cells of body clocks in all manner of organisms have found that timepieces tick away not just in the brain but in every bit of flesh, in every eye, kidney, and muscle, in every wing and thorax, in every leaf and petal.

We are ticking away in every part of our bodies. While the master clock in the suprachiasmatic nucleus probably oversees big rhythms in the mammalian body — sleep-wake cycles, body temperature, the secretion of hormones — the peripheral clocks scattered throughout our flesh may help specific parts of the body make what they need when they need it. The eye, for instance, may make different proteins at different times of day, each night renewing the tips of its rods used for night vision, and, at the end of each day, renewing the tips of cones used for color vision.

Michael Menaker, now at the University of Virginia, believes that the discomfort from shift work and jet lag may come about when these localized body clocks fall out of synchrony with the brain's clock and with each other. In a study of rats reported in the spring of 2000, Menaker and his colleagues simulated a six-hour time change for the rats by adjusting their exposure to cycles of dark and light. The rats' brain clocks shifted quickly to the new time, but the clocks in their lungs, livers, and muscles took much longer, a week to two weeks, to adjust to the new schedule.

The dependable little workings of these ubiquitous molecular clocks go to the root of our nature. By a neat trick of chemistry, clock genes measure a day through a self-winding mechanism. Two genes make two "daylight" proteins that turn on in the morning to make a second set of proteins, which accumulate during the day. When the second set reaches full strength, usually in the evening, it shuts off the activity of the daylight proteins in a negative feedback loop. The upshot is a genetic ticktock that reliably measures a twenty-four-hour day even without external cues, setting the body's daily rhythms. Light, even in minute amounts, works to shift or "entrain" the clock by raising or lowering the levels of key proteins, thereby delaying or jump-starting the next daily cycle.

While the details vary from algae to humans, this core mechanism of a feedback loop appears to lie at the heart of virtually all biological clocks, even those of photosynthetic bacteria, whose clock proteins have no resemblance to our own. Nature invented the design early and used it again and again. Some scientists suspect that this loop-the-loop routine, this cycling on and off, may be the only way to make a biological clock.

Nature loves a loop. I think of the owl tracking its mouse, translating the zigzag movement of prey into quick shifts of wing and tail in a tight feedback loop between mouse motion and owl flight; of water's charmed circle — evaporation, condensation, precipitation, evaporation — and the little internal circuit that turns the cellular craving for water into the precise notion of thirst. And, of course, the cosmic loops — dawn rounding into dawn; the moon shedding its shadow, as a snake does its skin, to be born again thirty days later; the sun swinging up from the horizon and down from solstice to solstice; the slow turn of the seasons.

These bigger rhythms are also enmeshed in our flesh. The *Oxford English Dictionary* tells me that the word "menstrual" is derived from the Latin *mensis,* or month, based on the word for "moon," reflecting the truth that the cycle of ovulation in all primates runs close to the moon's period. So, too, the seasonal high tides of the ocean, dictated by lunar cycles, may have their counterpart in the high tides of our bloodstream, as our blood pressure and temperature rise to a peak in spring, hence "spring fever." The composition and activities of our intestinal flora also vary with the seasons, as does our vulnerability to disease.

The biologist and philosopher Jacob Bronowski once wrote:

> Galileo is said to have discovered that a pendulum keeps nearly steady time by timing a swinging lamp by his pulse. All that Galileo or my doctor or anyone else discovered is not that either the pendulum or the pulse keeps steady time, but that they both keep the same time. Whatever their rhythm, they keep the same rhythm. We find the world regular as we find it beautiful, because we are in step with it.

We expect spring as we expect dawn, in the lens of the eye, in skin, muscle, and organ, as we expect fruit after flower, know in our bones that the hard ground will soften and breathe, the cold wind will dissolve in sweet warmth, time after time. The loop will hold.

To this old contract with Earth's predictable tick we owe the stability of our own little clocks. The details may have shifted. Nine hundred million years ago, the day was shorter by a few hours. But body clocks are so flexible that they've adjusted themselves to the Earth's slowing revolutions, just as the wizards of mechanical time have adjusted our atomic clocks.

I'm glad that the essential task of biological timekeeping was worked out long ago, the kinks largely eliminated. If it were left to me to set the daily, monthly, seasonal cycles of my body, I would get it all wrong, forget to turn up my hemoglobin, start my rod cells growing too late, confuse my lymphocytes. Or I might give up altogether and vote for an even keel. Imagine the disaster of that. Gardeners know that roots and bulbs must rest. Our metabolism, too, requires time out. Like our mechanical timekeepers — pendulum, quartz crystal, cesium atom — we are oscillators, from our unimaginably small tenth-of-a-second brain waves to the big beat of broader cycles, reproductive rhythms synchronized with the moon's period. Into our cells is built the two-part rhythm of planetary time. We never lose this basic dualism, even when we duck sunset and light up the night — which suggests why we feel lousy when we stay up past our evolutionary bedtimes or rise with our young in the dark hours before dawn to track the passage of a comet.

17

OF AGE

UP A HILL from my house is a cemetery with some old gravestones. I go there on bright autumn days to walk among the sun-warmed stones and collect leaves of russet, puce, umber, gold; to savor this time of year, everything flaming up, everything dying out. The French for autumn leaf is *feuille morte*.

Some of the new stones in this graveyard are carved and waiting with open-ended dates. I have considered buying a plot here so that I could visit my own grave. Occasionally I stretch out on the soil next to the Beasleys or the Maupins or the Livelys, kin buried with kin, and try to imagine what this is all about, conscious that under me are millions of living things, hordes of unnamed bacteria and worms ready to make good Hamlet's line. I look up at the fall leaves raining down from balding trees and think that "death" was only a word during the first half of my life, but is not necessarily so now.

There are a few sad graves here. Over there, next to the old caretaker's cottage, is a stone carved the year my mother was born and now barely legible:

> Unknown
> Young Man
> Found Dead
> on So. Ry. Train
> No. 4808
> May 7, 1929

A stone's throw from that, near a flowering crape myrtle, is a small granite monument to a baby girl who entered and departed the world on the same day. Season after season her creamy little stone bears a load of toys and pictures, a single pink rose, a miniature pumpkin, a Christmas angel, and, last winter, a white plush Valentine bear that froze solid in the snow.

Nearby, in the heart of the cemetery, are several stones dating from the 1800s, the graves of young women twenty years younger than I, most of whom probably died in childbirth. Not long ago people were often snatched away by plague or pleurisy or obstetrical failure well before age crept up on the body. Until the turn of the nineteenth century, life expectancy — the average of everyone's age at death — differed little from that estimated for the "average" Paleolithic man, who was expected to enjoy less than three decades. Even early in the twentieth century the typical American lived only forty-nine years. Since then, life expectancy has soared, chiefly because we have conquered high rates of infant mortality and many infectious diseases, at least in the developed world. Now members of our species are exploring the nether regions of life, well beyond the three score and ten allowed us by the Bible. We seem none too happy with what we find there.

My husband's mother, still a beauty at sixty-eight, says that the eyes she sees out with are hers. But the image they fix upon in the mirror, with its crow's-feet, slackening neck, and thinning hair, that is something different. The *I* remains, she says; it's this *Other* that changes. When I catch my reflection in the glass, I often see the face of my mother — the vertical frown mark just between the eyebrows, the set jaw, the lightly whiskered chin — and even, at times, the face of my grandmother, light skin stretched over cheekbone, softening underchin. What I see is a strange piece of flesh, the something other about me that is growing old, and I often turn away completely unnerved. On the wall in my friend's bathroom is a mirror, a present from her sister for her fifty-fifth birthday; it carries a little legend: "Mirror, mirror on the wall, you are your mother after all."

Age takes hold of us by surprise. Day by day we scarcely feel any perceptible change, don't take note of each new moment's slight metamorphosis. The slope runs down gently. It comes as a shock that the infinity of present instants we experience should bring about dramatic personal change. I know that I am literally not the same person I was a decade earlier, that the cells I possessed in youth have died away. Most of the cells in my body today were not present five or ten years ago — and some were not around yesterday. (Among the exceptions to this are the nerve cells in my brain, the cells in the lens of my eye, and the eggs in my ovaries, which are the same as those I possessed as an embryo.)

I know that most of my old cells have passed away in favor of fresh replacements. But I do not think of myself as a heap of cells, old *or* new. So I waver between the feeling that I'm stable and changeless and the certainty of transformation as understood by that glimpse in the mirror.

"In nature a repulsive caterpillar turns into a lovely butterfly," wrote Anton Chekov, "but with human beings it is the other way round: a lovely butterfly turns into a repulsive caterpillar." The verb "to age" literally means to grow older. But it carries with it the slur of decay, of slow, general decline. "Old" has the ring of insult, of second childishness. "The aged are fearful and hesitant," wrote Aristotle when young, "ill-natured . . . selfish, timid, cold . . . sorry for themselves." Being old made Yeats tired and furious: "An enemy has bound and twisted me so that although I can make plans and think better than ever, I can no longer carry out what I plan and think."

This is only half the story, of course. With age come the powerful mental benefits of experience and the gentling influence of perspective. "Great deeds are not done by strength or speed or physique," wrote Cicero. "They are the product of thought, character, and judgment. And far from diminishing, such qualities actually increase with age."

Still, for most of us, old age is the scary twin of sudden death. "The senses grow dull," said Pliny, "the limbs stiff, sight, hearing, legs, teeth, and even the organs of digestion move towards death faster than we." We are not alone. Most organisms with more than one cell decline over time. The skin sags, muscles weaken, kidney, liver, and lung decline, arteries harden, joints degrade, immune cells flag, neurons wither, eyes fog. So it goes until we end, sans teeth, sans eyes, sans taste, sans everything.

One spring day, thinking of aging, I took a book of family photographs from a shelf and found my mother and father, still young, leaning against their first car, an old Ford, turning, smiling, waving; my sisters and I years away from being born, no vows yet broken, no cancer yet diagnosed.

Wisdom holds that the lifespan of one's grandparents is a fair predictor of longevity. My gene pool is murky on this question. There's my father's father, dead of a stroke at seventy-four, and his wife — my grandmother — who did seem to shrink over the years, to have less physical substance in which to stand upright, but who grew sharper with advancing age, ever more quick-witted and independent, the sort of old lady who, nine decades into life, could live alone in her own apartment in the heart of New York, harry the local butcher for the best cut of beef, tell a wicked joke to a stranger on a park bench.

There's my great-grandfather on my mother's side, who outwitted nature (or at least the doctors), an organist diagnosed with stomach cancer at the age of forty-five and given six months to live, who carried on energetically with only half a stomach for another half-century — a warm, fun-loving man who, well into his nineties, threw stones at the windows of his sleeping great-grandchildren to rouse them for a game of baseball, in which he ran the bases. Then there's his granddaughter — my mother — who barely made it to the other side of fifty but whose wisdom was the wisdom of the elderly; and her mother — my Gagi — matriarch of our family, who lived to explore her eighties but only under the burden of insidious disease.

For forty years Gagi wrote poems to celebrate nearly every family occasion — birthdays, weddings, anniversaries, funerals — clever little couplets with tinkling rhymes or sweet, sober, mellifluous lines that neatly and precisely embraced her sentiments. The daughter of Clara Catherine Prince, poet and founder of the American Literary Association, Gagi was raised on the virtues of exactness in language, the inviolable laws of grammar, the dissecting of sentences at the blackboard, chalked phrases linked by a solid scaffold. Language used with clarity and precision held holy force for her, and she shuddered at any cheapening of it. She hovered above her children and grandchildren, intoning against split infinitives, drilling proper usage and grammar — lessons I failed to appreciate in my youth but later applied gratefully.

When Gagi reached her seventies, her poems slowed to a trickle, then dried up. By this time, my uncle had begun to help her with her financial affairs. Once, he noticed that she had written six checks on the same day in the same amount to the same name. The amount was $25, and the name was Jimmy Swaggart. Gagi was a devout Lutheran, as her father had been and his father before that. But she relished the music of Swaggart's voice, and, like the TV preacher, she spoke directly and often to God, sometimes praying aloud for a parking space. That day she had watched Swaggart's Bible show, hour after hour. Each time the preacher made his pitch for a donation, she took out her checkbook and made one, six in a row. It wasn't that she was swept away by Swaggart's message; she simply forgot what she had already done.

There were other examples of memory loss and confusion. Gagi would invite the family to supper and forget to prepare the meal. Some days she would roam the house endlessly, rummaging through drawers and closets looking for things she couldn't name. Once or twice in the dead of winter she wandered down the street in her nightgown, confused about who she was and where she was going.

Things got worse. Each day's events drained quickly into a dark hole of unremembered incidents. There seemed to be a sort of stuttering dialogue going on in Gagi's mind between the deep end of the past and an eroding present, with millions of events somehow lost or misplaced in the shoreface between. Sometimes she was unable to finish a sentence because she had forgotten its beginning, or she would ask the same question again and again as if it had just popped into her mind. Every so often she would snap back into her life and make ambitious plans, say, to start a garden or enroll in college. But then the valves would close once more and the awful funeral in her brain proceed. She experienced hallucinations and mood swings. I watched her self-composure, her normal flow of humor and sadness, dissolve sometimes into bewilderment, sometimes into agitation or petulant anger.

One day when I proposed that we stroll around the block, as the weather was nice, she looked at me with deep disgust. "Nice," she muttered under her breath. "*Nice.*" She glared. "What an insipid, foolish, EMPTY word!" She screamed at me to leave. Language, that last ordered force, I had squandered.

Gagi died when she was eighty-four. Although there was no autopsy, the diagnosis was Alzheimer's, the neurodegenerative disease that ravages the brain cell by cell, unraveling a lifetime of neurological connections.

Greek and Roman chroniclers wrote of the symptoms of memory lapse and disorientation. But Alzheimer's disease wasn't fully described until 1901, when Alois Alzheimer, a German physician, reported the case of a fifty-one-year-old woman who suffered from a devastating constellation of symptoms: memory loss, disorientation, paranoia, hallucinations. When his patient died, Alzheimer examined her brain tissue and found within its cells and cell fragments dense threadlike tangles and clumps of twisted and abnormal proteins now known as amyloid plaques.

Alzheimer's is the fourth leading cause of death in adults. It's twice as common in women as in men, and tends to run in families. For years it was thought to be caused by a virus or excessive exposure to aluminum or multiple head injuries. Now it looks as though genes are partly to blame. Three mutant genes are linked with the rare, early-onset types of the disease. A fourth, the so-called apoE4 gene on Chromosome 19, seems to play a role in the more common late-onset Alzheimer's, which strikes unpredictably at an advanced age. If you carry a double dose of the apoE4 gene, one from each parent, chances are about fifty-fifty that you'll get Alzheimer's some time after you're sixty-eight. The gene belongs to a large family called the apolipoprotein genes, which help store and transport cholesterol. ApoE genes come in three varieties, E2, E3, and E4. ApoE4 differs from the other two in only one nucleotide base out of nearly nine hundred, but it is this minute divergence that precipitates the onset of the disease, helping to build the plaques, the "tombstones" of Alzheimer's, that ultimately lead to the untimely death of brain cells.

Scientists using mice to model the effects of these mutant genes have made some unsettling discoveries. In the spring of 2000, researchers at the University of California reported that mice carrying the human form of apoE4 developed brain changes much like those in affected humans and suffered learning and memory lapses; they failed to locate and remember hidden platforms in water mazes. The effect was especially striking in aged apoE4 female mice. And here was spooky news: the learning deficit was apparent in mice only six months old. The faulty gene was doing mischief even to the young.

When I learned of this study, I was reminded of an odd phenomenon discovered in a group of Roman Catholic nuns in the 1990s. A team of scientists found a correlation between the nuns' writing style at an early age and the onset of dementia decades later. I remember reading about this study with grim fascination. While still novices in their twenties, the nuns had written autobiographical essays as part of their religious training. The scientists studied the essays, looking at their grammatical complexity and their "idea density," the number of ideas present in a single passage of text. Then they gave the now elderly nuns tests of memory, concentration, and other cognitive skills. They also studied the brains of fourteen nuns who had died since the study began. The nuns who had written essays with low idea density and simple sentences did poorly on the cognitive tests and showed signs of Alzheimer's after they died. But those whose essays had high idea density and lively, complex prose style did well on the tests, and none succumbed to Alzheimer's disease. This suggested the possibility that Alzheimer's may get its start early in life, as it apparently did for some of those mice with the apoE4 gene.

But good news about this Alzheimer's gene has emerged lately from studies of its evolution. The E4 type of apolipoprotein gene, it turns out, is the oldest of the three varieties; it's the ancestral gene inherited from our apelike ancestors. The two less harmful variants came into existence during the last 300,000 years and are becoming more and more common, taking stronger hold in each new generation, suggesting that our progeny down the line may be less susceptible to this form of the disease.

If Alzheimer's robs the elderly of youthful mind, cancer often robs the youthful of old age.

One thing about cancer, say the doctors, is that it comes on relatively slowly, allowing the patient to spend time with loved ones. In February of 1980, my mother was diagnosed with adenocarcinoma of the endocervix, a relatively "good" kind of cancer, with an 80 percent chance of survival.

She was fifty at the time. Lean when the cancer came, she ate less while it grew against the odds, thinning the walls of her skin, moving from her cervix into her lymph nodes, leaping to her liver, then making for her brain, gathering her into death cell by cell. I spent the month of May with her. She was too weak to sit, so she lay in the garden on a chaise longue while I hovered about the weeds. We talked in painful fits and starts. She had no regrets save one: leaving Beckie behind. She didn't want to burden her children. I tried to reassure her without making promises I couldn't keep.

In June she was hospitalized. In July she came home to die. On her last day, in the avid near-August heat, the air was like a wet towel. My mother lay in the half dark of the guest room, long-boned and thin in her last bed, drowsing in and out of consciousness while I swallowed goodbyes.

Powerful as she was, she was not powerful enough not to die before her time. And she was not, as I had imagined in childhood, like that old bird that flies out of its ashes.

Who's to say when aging begins? Dante believed that old age starts at forty-five; Hippocrates, at fifty-six. Charles Minot, an American anatomist of the early twentieth century, declared that aging begins at birth. Whenever it commences, aging does not normally, inevitably, include cancer — or even dementia. Not long ago, senility was considered only another float in aging's long parade of diminishment. It's known now that Alzheimer's disease is just that, a disease, not a normal consequence of cognitive aging, but a complex syndrome involving different factors, among them, genes.

In some way, though, Alzheimer's seems linked to the aging process; even in the normal elderly, decline in short-term memory tends to accelerate with age. In escaping the many illnesses and "normal" causes of death that took our ancestors — flu, the pox, measles — we're exploring a span of life that often carries a heavy burden of disease: heart trouble, osteoporosis, arthritis, cancer, dementia. This has lent an urgency to efforts aimed at unearthing aging's molecular roots.

In Greek mythology the three Fates controlled the thread of life: Klotho spun the thread, Lachesis measured it out, and Atropos cut it. Even Zeus was subject to their decisions.

Early scientists tended to blame aging on a single cause: the orbital circuit of Saturn, the body's loss of heat, the degeneration of the sexual glands. Hippocrates, who first likened the stages of human life to nature's four seasons, believed that old age was the result of an upset in the balance of the four humors: blood, phlegm, choler, and black bile. Autointoxication was the demon, said Paracelsus, a notion revived and refined by Élie Metchnikoff around the turn of the century. Metchnikoff blamed the ills of old age on the accumulation of toxins from putrefying bacterial organisms in the large intestines. So widely accepted was his theory that it inspired a fashion for colostomy and for a less radical antidote: establishing in the intestines a kind of bacillus flora that would stop the growth of debilitating sister organisms.

In the late eighteenth century, a German physician, Christoph Hufeland, proposed that every organism was endowed with a fixed store of vital energy that was used up over time. The "rate-of-living" theory of the 1920s embellished this idea. Energy was exhausted quickly or slowly depending on the owner's way of life. Smaller, faster-living creatures wore out sooner. This offered one answer to the question of why creatures tick through life at different speeds — a lesson I learned from the mice and gerbils of my childhood, loved and lost in quick succession. Mice usually live three years; gerbils, six; dogs, about fifteen; and Indian elephants, up to seventy. Fruit flies live for but a month or two.

However, not all life follows the rule that little animals burn up quickly and die young. Mutant dwarf mice live much longer than their normal counterparts. Some small mammals live up to seven times longer than the lifespan predicted by the rate-of-living theory. Some strange exceptions to the usual pattern of aging in humans, too, have hinted that something other than a simple wearing-out is going on.

Every so often, out of the millions in the human population, a child of twelve dies of a syndrome that looks like old age. This curious genetic disease, known as progeria, seems to step up the velocity of aging. The hair of a four-year-old whitens and grows sparse; the skin wrinkles and blotches with copper spots; the nose grows beaked; the chin, bony; the joints, stiff; the arteries, hard; the body, frail and hobbled. Most children stricken with the syndrome die of heart failure by the time they reach their teens. When the disease was first noted, in the nineteenth century, it suggested that aging was *not* necessarily linked to the abuses suffered through the passage of time, that there might be a mysterious clock in the body that ran independently of the chronological clock we use to count the passing years.

In 1965 the biologist Leonard Hayflick discovered that the body seemed to possess just such a clock: its individual cells. In a brilliant series of experiments on normal body cells grown in a test tube, Hayflick showed that after a certain number of divisions, cells suddenly stop dividing, grow old, and die. This number, known as the Hayflick limit, correlated with the lifespan of the creature. Cells from a Galápagos tortoise, which can live up to 175 years, divide 110 times, while cells from a mouse, which lives a mere three years, divide about fifteen times. Human cells curl up and die after fifty to seventy doublings. In all species, the older the donor, the fewer divisions its cells have left. It is as if an internal counting device were keeping a record of a cell's doublings and informing the cell of its age.

In the last decade a likely candidate for the aging clock within a cell has turned up in an unlikely place: at the very tip of each of our chromosomes. As early as the 1930s, Barbara McClintock had proposed that chromosomes must have something at their ends that keeps them from falling to pieces or sticking together or otherwise misbehaving during cell division. That something is a telomere, a bit of what was once considered junk DNA, a repetitive stutter of base pairs that forms a little loopy cap at the end of a chromosome, protecting its store of genes. Virtually all organisms with nucleated cells — yeast, frogs, mice, humans — have telomeres.

Every time a cell divides, its telomeres shrink a little, losing base pairs. After a certain number of cell doublings, the telomeres get too short to bend into a loop, and the broken end of the chromosome signals the cell to stop dividing and to proceed in an aging mode. It is as if every time you knotted your tie, a bit of it frayed, until after a certain number of knottings, the tie completely unraveled. The telomeres of an octogenarian are, on average, far shorter than those of a child. Except a child with progeria, whose telomeres resemble those of the elderly.

If a telomere is a clock of sorts, an enzyme called telomerase is the key for rewinding it. The enzyme lengthens telomeres by adding bases to DNA, little stretches of buffering "junk." While all human cells possess the gene that makes the enzyme, in most normal body cells it's silent. Cells that make the enzyme — most notably eggs and sperm, and also those that revitalize blood and skin — are immortal. Cells that don't make it are doomed to age. (Cancer cells somehow reactivate their telomerase and go on youthfully dividing.)

It's tempting to imagine that my joints stiffen, my cheeks wrinkle, my eyesight fails, all because of this chromosomal clock and its chemical winder — Klotho, Lachesis, and Atropos rolled into one. But trying to catch aging in the single net of telomeres is as misguided as blaming an imbalance of the four humors. Like most biological phenomena, you can't starve the process into simplicity.

Next to my file of biological surprises is one labeled *Genes For.* Among the items is a report of two independent studies fingering a single gene "for" the love of novelty. The report claims that people who are intensely curious, adventurous, excitable, extravagant, impulsive — "novelty seekers" — tend to have a longer version of this gene than do the more staid among us.

The gene codes for a receptor that allows the brain to respond to dopamine, a chemical signal strongly linked to pleasure-seeking behavior. (The findings have since been disputed by scientists who found no signs of the link in Finnish men; others noted that the gene was far more frequent among people in the Middle East than those in East Asia and therefore an unlikely candidate for such a widespread trait.) Following on the heels of this find was the discovery of the gene for anxiety or worry, a shortened version of a gene on Chromosome 17.

Such announcements about a single gene linked with a particular behavioral trait are sprouting like spring leaves on a tree. "It's easy in all the excitement to confuse a genetic tendency toward a trait with the trait itself," the biologist Ruth Hubbard told me, "to suppose that a single gene that contributes to intelligence, say, or sexual preference actually specifies that trait. It doesn't. Attributing biological process or behavioral traits to individual genes is about as meaningful as saying 'blood will tell.' The way in which genes may or may not contribute to personality traits and behavior is multifarious and complex; linear, reductive, simplistic attributions just won't do."

A professor of biology emerita at Harvard, Hubbard once studied vision in frogs, squid, and cows — specifically, the architecture of such visual pigments as rhodopsin — and has more recently become known for her work on ethics in science and on the public perceptions of genetics and molecular biology. She has a slight tremor in her hands, which she says *is* purely genetic; her mother had it, and so did her grandmother.

"It's reasonable to think of genes as recipes," she said, "for a custard, say." She laughed. "I have enormous respect for custard, almost as much as for personality. But the key is what happens from recipe to dish."

It isn't enough to know the ingredients. A single gene almost never acts alone, but interacts with other genes in surprising ways. While one gene is firing off a signal for me to dive from the edge of that quarry cliff into a deep blue water hole — a molecular version of Poe's "imp of the perverse" — a dozen others may be pumping out the neurochemical equivalent of my overprotective grandmother. All genes are embedded in the layered biochemical environment of cell, organism, outside world.

This is true as well of so-called disease genes. In something like half of all human tumors are cells with mutations in the p53 gene, that "guardian of the genome" that looks out for wayward growing cells, the founders of cancer. Most cancer treatments are thought to work by alerting p53, but the gene itself is vulnerable to damage from ultraviolet light, toxic chemicals in cigarette smoke, and other environmental carcinogens. When it's mutated, the p53 protein misfolds, losing its ability to bind to DNA, to whisper its instructions to a roguish cell, and the road is open to cancer. A mouse with a disabled p53 gene will develop tumors within a few months. A human with a mutant form of the gene may suffer from multiple cancers, a hereditary disease known as Li-Fraumeni syndrome.

But a faulty p53 gene is by no means the single fuse for cancer. Several genes, as many as ten or twenty, must be changed to transform a normal, obedient cell into one devoted to morbid excess. The genes in question, so-called proto-oncogenes, are altogether normal and necessary for life; it's when they're mutated that they can turn a cell cancerous. The breast cancer genes, BRCA1 and BRCA2, are expressed in normal cells and are important to their workings, but their disruption predisposes a woman to cancer of the breast. Whether or not she gets cancer depends on a slew of other factors: which hormones, carcinogens, and viruses she has been exposed to over life, and the variations in certain other genes, such as those that enable her body to detoxify chemicals. Some versions of these genes allow the body to act against toxins rapidly; others, at a more sluggish pace.

The response of genes to their surroundings is nearly as complex as weather and just as difficult to predict. The only way to grasp a gene's meaning, says Hubbard, is to observe its activity throughout life in its full chemical and physical context.

It's the same with aging. In no single gene or cellular mechanism can be found the one instrument that ages us. Telomeres aren't the whole story. For one thing, brain cells and certain muscle cells don't divide during adult life and so don't suffer the effects of shortened telomeres; no one knows what controls aging in these cells. And while telomeres do clearly shape the aging of many types of human cells in a petri dish, it's still not clear how they affect aging of the whole organism.

Certainly there's no gene "for" aging. A team of Japanese researchers found a gene in mice that, when mutated, causes a syndrome that resembles human aging. The gene, which normally makes a kind of protein found in the cell membrane, can, when it's disabled, cause age-related disorders — arteriosclerosis, osteoporosis, skin changes, and shortened lifespan — earning it the name *klotho*. But this is only one gene in a bevy of genetic players.

Scientists from the Scripps Research Institute recently took a close look at the variety of genes that change with age in cells known as fibroblasts, which build skin and connective tissue. They compared the genes of young, middle-aged, and old people and also of people who have progeria. Out of some 6,300 genes they examined, 61 changed with age within these cells, both in the normal aging process and in progeria (suggesting that the disease indeed may be an accelerated form of aging). Among the changed genes were some involved in the making of collagen and other proteins in skin and tissues; some linked with inflammation, a process implicated in several of the diseases associated with aging — heart disease, arthritis, even Alzheimer's; and others that help to control cell division. It is these last that the scientists find most intriguing, a possible key to the myriad physical expressions of aging. Defects in these cell-division genes can make chromosomes unstable, triggering misbehavior in other genes that gets worse with each round of cell division, eventually affecting a mosaic of cells: immune cells, skin cells, neurons, muscle cells. Aging, then, may be linked with this fundamental genetic mishandling of cell division.

Certain genes are thought to possess the power to slow aging and add years to life. Fruit flies with a mutant *methuselah* gene live far longer than ordinary fruit flies. A little genetic manipulation can radically extend the lifespan of *C. elegans*. Scientists have added days, even weeks, to the normal twenty-day lifespan of the worm, sometimes by selectively breeding for longevity; sometimes by mutating a single gene. Change a worm's *daf-2* gene, and you can double its days. No comparable gene has yet been found in humans, but some biologists believe it may turn up soon.

Another factor in aging is our rather imperfect body chemistry. In the course of normal living, the body makes defective molecules and toxic waste products that sabotage cells. The simple sugar glucose, which fuels us, is sticky. In the cell it has a nasty habit of latching on to molecules and disabling them. Collagen and elastin, when gummed up this way, make the eye lens less elastic and flexible, and connective tissue in the joints stiffer.

There's more. In processing oxygen, cells make oxygen-free radicals, molecules with a single unpaired electron. To balance their structure, free radicals steal electrons and atoms from other molecules and, in so doing, may damage DNA and proteins and literally melt away parts of a cell's membrane. That creates more free radicals, which damage other molecules, in a runaway chain reaction. Radiation from sunlight can have the same effect. Bruce Ames, a biochemist at the University of California, has estimated that oxygen radicals damage the DNA in a single human cell something like ten thousand times a day.

There's also speculation that our mitochondria may contribute to aging, either by releasing reactive oxygen molecules or harmful proteins that set in motion apoptosis, or by falling apart themselves and thereby robbing a cell of its energy.

That we don't disintegrate instantly in the face of this bombardment, but manage to blunder along for decades, is largely thanks to our sophisticated systems of self-replenishment and self-repair, and to a bewildering array of substances made by our cells that soak up free radicals. Several of the newly found genes that lengthen the lives of yeast, worms, fruit flies, and mice make these so-called antioxidant proteins. A mutant mouse with a single mutated gene that heightens its resistance to agents causing oxidative damage lives almost a third longer than a normal mouse.

Wounds heal, bones mend, damaged DNA zips itself back into shape. Nevertheless, despite the impressive, insistent energy that sings us back together, we still tend to fall to pieces over time.

Could things ever be otherwise?

Once when I was exploring a fossil bed in far northeastern China, a landscape that, more than a hundred million years ago, was a lake, I found a beautiful fossil of a mayfly larva, *Ephemeroptera,* imprinted on a brittle flag of volcanic rock. To my amazement, I found another larva, then another, and another, dozens within reach. I should not have been surprised to see the mayflies in such numbers. As I child, I found the insects thick in the waters of neighborhood streams, where they passed their wormlike larval state. As their name implies, mayflies metamorphose and emerge at the apogee of spring and then hover through a curiously brief flight of life. Lacking mouthparts or intestines, they don't feed at all; they mate, lay their eggs in huge numbers, and die, all within a few hours.

Other creatures pose a similar riddle of brief existence, dying right after they reproduce without going through any phase of degeneration. In the mating season, Pacific salmon in their prime move up a stream against a rush of white water, traveling hundreds of miles to spawn in the pools where they were born. A few days later they are fodder on the streambed, their bodies rife with ulcers and teeming with infection. Octopuses and eels, too, flame out and die young.

Alligators, turtles, sharks, lobsters, rockfish, and sturgeons don't senesce but go on springily to advanced ages. Certain kinds of tube worms lodged in the dark depths of the sea can live for more than two centuries. Lobsters up to a hundred years old snap their claws as briskly as young ones. A fifty-year-old mud turtle looks and acts like a turtle half its age. Most single-celled organisms — bacteria, slime molds, diatoms — remain eternally youthful. In what way did evolution tap these organisms for perennial youth? Why should anything go wrong in our bodies with the passage of time? Why should we decay?

One theory holds that creatures that live life without senescing are not cast in a new revolutionary mold; rather, just the opposite. They represent an old way of doing things. Aging — the sorry tendency to weave through long life while steadily falling to pieces — began as an innovation, an exception, something new.

Just why it came about has bedeviled scientists. The evolutionary biologist George C. Williams has proposed that senescence and slow death are the devil's half of an evolutionary bargain, the upshot of a set of trade-offs between features that are useful to a young organism and those that serve an older one, a concept known as "negative pleiotropy." Take the makeup of a knee or an ankle joint. There may be a trade between a strong, light material that wears out quickly and a material that is heavy and taxing but more durable. Or consider the possible swap between a chemical process that makes strong, vibrant proteins but pollutes the body over time, and one that makes inferior proteins but is clean and harmless. The feature that strengthens the young and fertile — the strong, light, short-lived joint material or the vibrant, polluting protein — tends to win out, even if it causes mischief at the tail end of life.

Nature strives to preserve eggs and sperm over other sorts of body cells. Mutant genes with harmful effects on young people, such as those responsible for certain types of progeria, are rare because natural selection eliminates them. But genes of similar power, with effects limited to people over the age of ninety, may go on and on. The debilitating aspects of aging occur after we reproduce, so there's no pressure from natural selection to weed out the guilty genes. They pile up in the human genome over evolutionary time, eventually affecting not just one or two organs but the whole body. The result is synchronized frailty and decrepitude. Once we've reproduced and ensured the survival of our offspring, it matters hardly at all whether — or how — we collapse inward with age.

When I was little I thought the silvery lustrous ooze left by the slug as it crept across my mother's patio was its own leaking flesh, that the more journeys it took and the farther it went, the more self it lost, dwindling down with age until it simply smeared itself out of being.

The truth about aging is somewhere inside us, no doubt, but the trick has been to get the secret out, and, once it is revealed, to be sure that it is read correctly. We know it's a tangled and untidy affair, having to do with genes, telomeres, hormones, cells, and the imperfections of our body's chemistry. We know that simple good habits — exercise, coupled with healthy eating — can slow the progress of aging and even reverse its effects, build muscle, strengthen bone, and spur the growth of new nerve cells, improving learning and memory, at least in mice. Still, we seek the one fountain of youth, the single key that will turn back the aging clock.

We are already defying the odds. We live far longer than would be predicted for a mammal of our body size. Compared with fruit flies and mayflies, we are as everlasting as Methuselah. But our view is necessarily parochial. To us, a lifetime can seem but a breath. One day I'm a girl with my front teeth missing; the next, a woman of middle years whose mother died at about her age.

Present always has been the keen hope that we might somehow postpone joining our ancestors. To extend the number of his years, the elderly King David was advised to lie between two virgins. Medieval alchemists believed gold's absorption into the body would prolong life. The year one dreamer discovered the Americas, another, Pope Innocent VIII, drank the blood of three young donors in the hope of adding a few years to life; he died shortly after. In the nineteenth century, a seventy-two-year-old French professor, Charles-Édouard Brown-Séquard, attempted to regain his youth by injecting himself with extracts of dog testicles, in the belief that a deficiency of male hormones was at the root of aging. We are only as old as our glands, wrote Eugene Steinach of Vienna in the early 1900s; he recommended vasectomy. Yeats was Steinached at the age of sixty-nine, after the loss of Lady Gregory halted his writing. Charlie Chaplin and Winston Churchill took injections of fetal lamb cells in vain efforts to defy the passage of time.

Lately, scientists seeking clues to aging and longevity have tinkered with resetting the telomere clock to lengthen a cell's life. They have found a way to flick on the telomerase gene in cells so that they can quickly build their telomeres to youthful length and go on dividing up to ninety times. But turning on telomerase may come at a high price: making cells immortal makes them more likely to become cancerous.

Restricting calories has been offered as a way to delay death. The lifespan of mice and rats may be greatly lengthened by severely limiting their intake of calories, which seems to step up DNA repair, decrease damage from free radicals, and preserve the immune system. By such deprivation the days of the water flea and of the bowl and doily spider may be doubled. Indeed, studies suggest that reducing by half the food intake of all sorts of animals extends lifespan. Scientists studying yeast recently pinpointed the way caloric restriction might work to lengthen life — by means of genes that switch on when food is scarce, silencing other genes and thereby helping to limit wasteful activity in a cell.

So, too, limiting the number of one's offspring may increase the measure of our days. Experiments with fruit flies show that those individuals with fewer offspring live longer. A recent study of the genealogical data from twelve hundred years of British aristocracy suggests that this correlation may also apply to humans.

I have not yet reached that age when I might lie awake and stare into the dark while the clock strokes, the small worries of the day transmuting into the deep fear of dying. But my midlife view is this: If semistarvation and childlessness is the way to longevity, I would rather flame out early like a salmon or an eel or burn to umber like a maple in the last days of fall. I vote for the delight of pie, the sweet pleasure of children, and a regular descent into a dark, soft, willing earth, waiting there to take what it has given.

18

SWEET MYSTERY

WATCHING LIFE, it's easy to spot the constancies. The push to birth is a dead giveaway, the urge to break or squeeze toward daylight through shells, seeds, vaginal tracts. So is the hunger for growth, for dividing and multiplying, for clumps of cells, masses of eggs, milky clouds of larvae. So, too, the tendency to separate, to make boundaries, membranes, skin, but also to join, to merge, knot, and pool, to flock and to swarm. Likewise, the impulse to fidget among creatures — to tremble, blink, shimmer, wobble, shiver, flex, and clench — and to hold on, to grip with hooks, suckers, and little flicks of keratin. And the call to voice, to signal, hoot, howl, hiss, chirp, bark, wail. Pervasive in life is the propensity to breathe, eat, digest, excrete, copulate, collaborate, and conspire, to suffer aging and death.

We did not have to wait for modern biology to tell us that we are akin to other creatures. It was probably our first great thought, with our totem systems and animal folktales. Across the illusions of form there is kinship, "a principle of continuity that runs through the scale of structure in living things," wrote Aristotle. "And so, little by little, by imperceptible steps, does Nature make the passage from the plant, through animal, to Man."

I grew up with the idea that this natural passage began in deepest time in some warm, placid little pond on the skin of the planet. Somewhere, the wizardry of sunlight striking a tepid pool vaulted a lucky collection of molecules over the hump into primitive life. But evidence has lately surfaced that points to another possible place for the origin of life: a hot, swirling sulfurous cradle straight out of hell — the hydrothermal vents at the bottom of the sea.

Such places first came into view twenty-five years ago from the portholes of the little research submersible Alvin, exploring the ocean depths two miles down near the Galápagos Islands. In spots were chimneys in the earth's crust, "black smokers," foul night worlds of blistering heat and intense pressure, where magma and superheated water laden with toxic hydrogen sulfide welled up in plumes to meet the icy waters of the abyssal sea. Here was a place where life seemed unimaginable, only a little less hostile than the black reaches of interplanetary space. But here, violating all our assumptions about nature, was wild, luxuriant life — giant clams, hairy gastropods, six-foot tubeworms in showy crimson clusters, and blizzards of bizarre microbes thriving in impossible heat, darkness, and pressure. Among the microbes were colonies of bacteria never before seen, including some that looked to be descendants of ancient archaebacteria, perhaps the oldest ancestors of us all.

Evidence of just such ancient microbial life turned up in the hot summer of the new millennium, when Australian researchers discovered lifelike microscopic filaments, "twining and twisting in different directions," in deep-sea volcanic rocks more than 3.2 billion years old.

We think we have life down; we think we understand all the conditions of its existence; and then along comes an upstart bacterium, live or fossilized, to tweak our theories or teach us something new.

Although deep-sea vents — rich in minerals, protected from the comets and asteroids that bombarded early Earth — are attractive sites for the origin of life, there are some who still argue for a cooler cradle, Darwin's little dream pond or cold ocean or even a lattice of ice. But one thing seems clear: cool pool or Hadean vent, before life could pop forth in any form, it had to be assured of nature's stability. That assurance was water.

If there is common ground on this planet, it is contained in this liquid, Melville's "sweet mystery," the universal element in which chemicals can combine to make metabolisms. Water is found in spots elsewhere in the universe. The masters of astronomy have spied molecules of water in clouds of interstellar gas and in the spectra of red giant stars. They've seen signs of it in cool brown dwarfs, beneath the thin crust of a Jovian moon, even in sunspot umbrae. But on Earth it is everywhere — in the great bulging blue eye of the sea, clinging to the backs of beetles, oozing secretively underground beneath the ice-locked poles, and at temperatures of 1000°C. or more beneath the planet's skin.

Whether Earth's water first came from a rain of comets, as some suggest, or arose from clouds of vapor produced by great volcanic eruptions, nearly all of it was created in the first few hundred million years of the planet's history and persists today, flowing now into glacier, now into the particles of mist hanging over sea cliffs, now arraying itself out of turbulent air in the unique lacework of a snowflake or flushing through the bloodlines of our children, puddling in their cells, prescribing the lives of all organisms in ways we've just begun to grasp.

The summer I was ten, I got what I'd asked for, a chance to swim with my family in the clear, swift waters of a New England river, a summer retreat renowned as a place of magic. The river was cold and deep, with a strong current. Due to some trick of topography, mineral, and flow, its waters had worn the bedrock smooth as glass so that a child could ride the rocks as if they were a waterslide, though she was warned to avoid the treacherous side streamlets, with their drops and holes. Drowning in the river's peculiar undercurrents and whirlpools was not unheard-of.

The day was warm and bright. I shed my clothes and eased into the river's fringes, felt the shock of cold in the shallows along the bank while I gathered my nerve, then pushed off and floated into the swift current of the central channel. Borne along, a leaf in a stream, I drifted feet first and face to the sky, sensing the smooth rock beneath me, the hard face of the old Earth over which I was sailing. Suspended in the gleaming medium, I felt buoyant and graceful, momentarily released from the gravity of space and time, freed from the confines of flesh. I turned over on my belly and porpoised, face first, in the torrent. Then, in a panicky instant, I lost the flow, slipped away from the main channel into a streamlet. I was sucked down, dragged under, and held half a minute before being released, head first, into a midstream boulder. Up from the whirl I popped — shaken, burbling, and choking — with a bloody mouth and a cracked front tooth.

Now water pursues me. Waking or sleeping, I dream of it, of floating peacefully, face to the stars in the dark, calm flow of the river I grew up on, or face down, struggling to right myself in its muddy flood. I am drawn to trickling springs and limpid aquatic gardens, the damp evocative rot of marshes, the splash of ornamental runnels, the shore's heavy surf.

Though I get seasick when I'm out in the middle of the ocean (a polygenic disposition that runs in families, which makes me wonder at the fortitude of all my German and Russian relatives on their transoceanic journeys), still I love the vast open waters of the sea.

In the eighteenth century a French diplomat proposed that man's immediate ancestors were aquatic people. They spent part of their life underwater and often had fins instead of feet, scales instead of bare skin. The diplomat claimed that "about ninety such creatures had been sighted," wrote Evan S. Connell in *The White Lantern*. "Several females were delivered to the king of Portugal who, wanting to preserve these curious beings, graciously allowed them to spend three hours a day in the sea, secured by a long line. . . . It is said that they submerged at once and never came up for air."

Two hundred years later, zoologists at Oxford revived the idea with the serious suggestion that our ancestors were semiaquatic apes, that this aquatic phase in our evolution might account for some of our more curious features. The writer Elaine Morgan has defended this notion with courage. We are unique among land animals in having a "dive reflex," she points out, and, like dolphins and hippos, we are hairless mammals. Our bodies, like those of marine mammals, have an insulating layer of subcutaneous fat bonded to skin, which makes us both buoyant and streamlined. Yet most scientists discount the liquid roots of these strange traits and say the cradle of human evolution was not a body of water, but a broad sweep of grassland.

Whatever the habits of my more recent relatives, my earliest ancestors were born in water, and when they crept up into the throttling air, they brought along the liquid inside their cells. As Loren Eiseley once said, all of the differences between living forms have been achieved only by the elaboration of devices for maintaining that precious liquidity without which cells cannot live and grow: "Not for nothing has the composition of mammalian blood led to our description as 'walking sacks of sea water.'"

When my mother was cremated, I wondered how the bulk of her — skin and bone and muscle, brain and heart and strong mothering hands — could, even in her last emaciated state, turn into so few ounces of ash. I had not considered her constitution. We are liquid beings, 70 percent water, like the Earth. The proportion is the same in a bacterial cell as in a human cell, a likeness I find extraordinary, as if every organism had somehow swallowed a planetary ratio.

For humans, at least, lack of water is the taste of death. A drop of 2 percent in our body fluid immediately manifests itself as thirst; a 5 percent loss induces hallucinations; and a loss of 12 percent will kill us. Water is the medium of all our chemistry. Neither root nor flesh can absorb nutrients unless they are dissolved in water. Only by water squeezing through the wall of the kidneys, the sweat glands, and the lungs can we excrete our own poisons.

In the car one summer, in the middle of a long, devastating drought that burned grass and flowers, put cows up for early sale, pinched my neighbor's crops, and filled the radio broadcasts with dire warnings of low reservoirs and dry wells, Elinor, my four-year-old daughter, piped up from the back seat in a demanding voice, "Who invented water?"

Imagine the task. Take two hydrogen atoms and a single oxygen atom and make them into a molecule shaped like a V. Make the angle between the arms 104 degrees and the distances between the atoms — the dashes in H-O-H — precisely .095718 nanometer. Make the molecule conservative and self-loving by giving it an odd electrical asymmetry, clustering the electrons near the oxygen atom, allowing one molecule to bond easily with another so that rivers, lakes, and oceans hold together, so that water remains liquid at room temperature when it should be gas, so that my metabolism, the basic business of my bodily living, does not bring on a temperature that would set my bones afire.

Now, make the number of molecules present in a mere half-ounce of water a number of such magnitude that it's nearly impossible to grasp — even if you tell me it equals the number of teaspoons it would take to empty or fill the Pacific Ocean. Make those zillions of molecules restless and mobile, dancing in short-lived flickering clusters, linking, separating, linking again, swapping partners a hundred billion times a second, so that their very motion diminishes the forces that link other atoms, freeing them to combine chemically with free-floating atoms, so that water is a great solvent, able to weather rock, strip mountains, dissolve and transport the substances of life.

Water finds its way inside our molecules and leaves its strange stamp on body chemistry. DNA and RNA love the liquid and will happily mix with it. The lipids that make up the outer membranes of our cells are of divided feeling, with heads that love water and fatty tails that abhor it, a lucky antipathy that defines each cell's boundary and gives all of us definition. Of two minds, too, are the different varieties of amino acids. Half like the liquid and cluster on the surface of their folded-up protein chain so as to be exposed to the brew; the other half dislike it and tend to congregate on the inside of the folded-up molecule, where they stay dry.

When the biochemist Leslie Kuhn went looking for water in the dark microspace of proteins, she discovered a neat surprise. Scientists had known that up to 77 percent of protein crystals consist of water. Like proteins themselves, though, the water molecules within a protein crystal cannot be seen even with the most powerful microscopes. So Kuhn worked at seeing them indirectly by means of calculations and formulae. She found that water nestles into the deep clefts and grooves of a protein's landscape, those key binding sites where most target molecules attach to a protein. "Nestles" is perhaps not the right word: the water molecules hop in and out of grooves at a boggling rate, more than a million times in a thousandth of a second, and yet they serve to stabilize the surface topography of the protein.

This is true for proteins in virtually all organisms — yeast, cows, humans. Kuhn and her colleagues at Michigan State University looked at the water tucked into similar enzymes in a range of creatures and found that "groove" water is conserved across species. Without this water, fidgety though it may be, the protein's folds might actually squeeze its grooves right out of existence. Water keeps the door open so that the amino acids deep in the grooves are accessible to target molecules. Without water to urge folding in the first place and to hold open the sites of communication, proteins would fail at their marvelous sweep of activities, fail to ferry iron, to bind sperm with egg, to crystallize the lens of an eye, to direct an embryo toward bilateral symmetry and toward birth.

Not long ago, news of a bold new theory describing the liquid origins of our earliest ancestor gave me a great jolt of pleasure. The oldest shared ancestor of all living things, the universal ancestor, proposed Carl Woese, was no single entity, no particular organism, but, rather, "a loosely knit, diverse conglomeration of primitive cells that evolved as a unit."

Woese, an evolutionary microbiologist at the University of Illinois, envisages this: billions of years ago, in a boiling cauldron at the bottom of the sea, a group of proto-beings rubbed together, perhaps in a sulfurous swirl of magma and seawater. The members of this prehistoric commune may have differed from one another; they may even have had different genetic codes. Under harsh and changing conditions, the organisms that endured were those that could make good use of their neighbors' genes to adapt — swapping them, linking metabolisms, trading molecular products, toxins, nutrients, simple waste. In this way, the organisms able to read DNA-based genes outlasted the rest, and bit by bit, cluster by cluster, the community gave rise to cells that fully mastered the secret of faithful replication and burning to a purpose. In a short time they became miraculously complex, working their way into archaea, bacteria, eukaryotes — all the lineages of life.

Here's a poker to stir the fires. Just as I've settled into the idea that I'm an edifice, a colony of perhaps sixty trillion cells that have bound themselves to make one being, and that each one of those cells is in itself a kind of colony — a population composed of mitochondria, relicts of ancient cellular mergers that now interact within a membrane — along comes this new theory that takes the notion a giant step farther. The first proto-being, whatever it was, the Ur-ancestor of us all, did not just arise *in* a pool, it *was* a pool.

Wherever we look, as we dissect the workings of beetle, bacterium, squid, fish, or mentally crawl back through the long vein of geological history to the birth of life — we find hints of hidden connections, molecular links that huddle and lump us with the rest of life. This doesn't change the stories I grew up on, but it does change the way I hear them.

We can't feel as separate from other creatures as we did a decade ago. Though in our imagination we might continue to plant ourselves above the circle of the moon, and bring the sky down beneath our feet, now more than ever we see ourselves passing by in the face of a cephalopod, the nose of the vole, in the amused little eyes staring at us from low on the evolutionary tree.

The biologist Robert Miller once wrote:

It took nature more than a billion years to develop a good worm, meaning one that has specialized organs for digestion, respiration, circulation of the blood and excretion of wastes. . . . [Worms] also developed segmentation or reduplication of parts, permitting increase in size with completely coordinated function. Contemporary architects call this modular construction. It is found in man in the spinal column, in the segmental arrangement of spinal nerves, and in some other features that are especially prominent during embryonic development.

I will always love an earthworm far better than an exon, would rather tease out the centipede's tiny reduplicated segments than those of a hemoglobin gene. I am far more passionate about the hognose snake than about the molecules that shape its fine limbless blueprint, would rather split hairs to distinguish a Cape May from a Blackburnian warbler than finger the various members of the G-protein signaling family. At heart I am skeptical of the "nothing but genes" school of thinking, the bleak summary of life as DNA's way of making more DNA. It is birds I choose to name and watch and imitate. To lapse into jargon, my deepest feeling is for the warm, thumping, keening, crowing phenotype.

But the strange parallel planet below the field of vision has its own blood-and-nerve vitality, and teasing life apart cell by cell, gene by gene, doesn't diminish the wonder. On the contrary, it brings to light new mysteries. "The universe is not just queerer than we suppose," said J.B.S. Haldane, "but even queerer than we can suppose." If molecular biology has revealed anything in the last decade or so, it is the tendency in nature toward surprise, toward the unpredictable, toward making what we know point an inquiring digit toward what we don't know.

In one of his lectures, the late physicist Richard Feynman invited us to imagine that we were blind to rainbows but knew of their existence only through measurements. Would the mathematics of a rainbow, purely considered, be beautiful? Does the essence of a thing ever lie in the assemblage of its microscopic parts?

I don't say that there is nothing in humans that was not first in amoebas. That is to stay too long at the spark. Thousands of millions of years have got us where we are, to learning, thought, creative discovery, the power to name and identify, to fashion from the noisy void a lexicon for the fish of the sea and the genes of the body. We are perhaps unique among all creatures in the sense that we can talk about evolution and understand it. We can frame bold new theories, arrange a combination of words that explains DNA, and wonder — only four years into life — who invented water. In the highest, hardest part of our minds, we have the ability to worry that other species on this planet are going extinct with a million things unsaid, and that we may not be able to do without these lost relatives, just as the proteins and threads of nucleic acids in our bodies can't do without water.

Ultimately it may be possible to explain all life in biological terms, to reduce all the complexities of cephalopod intelligence, bird song, poetry, to the interplay of genes and proteins. But we are a long, long way from this. We know better than to claim completeness of the picture. We know that claims for absolute certainty are unwise. The evolutionary tree of life, if you can call it a tree, is defying our arboreal expectations, looking more like a bush with branches that merge and split and merge again, hopelessly intertwining. The "base" of the tree itself is beginning to look more like a web, like Darwin's mysterious tangled bank. A firm answer to the origins of RNA, DNA, proteins, life still eludes even the nimblest imaginations. We are not even sure whether the origin of life was a slow, gradual growth over eons or a singular molecular POP out of a wormhole in the universe.

Innumerable riddles remain wrapped in that blind bug of a term, DNA. We have mastered an extraordinary task in spelling out the genetic alphabets of creatures from *E. coli* to *Homo sapiens*. But we are only beginning to grasp their meaning, to learn what individual genes do, how they work together, how they're changing and remaining the same. We're still struggling to fathom how much of a genome is useful, how much is junk. We still don't have the ghost of an idea why a particular gene may do much important work in one creature but sit unobtrusively in the corner of another, like an unplayed note on a score.

In nearly every genome studied to date, in creatures from bacteria to *Drosophila*, a quarter to a third of the genes bear no resemblance to anything we've seen before. "Orphan" genes of "unknown function" with "no obvious homologues": in such phrases we get the feel of the unknown that never quite separates itself from the inner workings of life. Such orphan genes make up a vast *terra incognita,* offering the possibility of new lessons in likeness, difference, order, being, seeming, and by themselves good reason for humility.

We are still puzzling over the 2 percent difference in DNA that separates chimps from our own tribe. That small percentage holds fourteen million possible nucleotide differences, no trifling number, yet less than those between a chimp and a gorilla.

We are beginning to get a handle on the notion that humans, all humans, are essentially alike — more alike than are individual members of other species. Our differences are a matter of slight variations in a few genes that account for such superficial traits as skin color, slim differences that have been the cause of great suffering.

We are grappling with how to use the genetic wisdom we've gained, whether it's wise to lay bare the traits in our genes, the predisposition to this disease or that ability. As I finish this book, geneticists are making discoveries that have personal implications for my family. Many heretofore unexplained cases of mental retardation are caused by slight chemical rearrangements at the tips of chromosomes, a pattern of chemical reshuffling so subtle that it eludes conventional genetic screening. It is now known that this reshuffling tends to run in families — knowledge I'm glad I lacked during my pregnancies and wonder how my children will handle.

We're still pondering a mystery raised more than a decade ago by Barbara McClintock: the extent of knowledge a cell has of itself, and how it may use this knowledge in a "thoughtful" manner in times of stress to initiate its own restructuring and renovation. New curiosities in this arena are surfacing almost daily. For instance, the discovery that heat-shock protein 90, a protein ubiquitous in nearly all forms of life, allows genes to store mutations — instructions for sudden, radical change in body shape — and to haul them out in case of a drastic change in environment. (Heat-shock proteins, which help restore the folding of proteins disrupted by high heat, may be part of the housekeeping chemistry that evolved early in the history of life in the hazardous environment near hydrothermal systems.) In our cells, then, are reservoirs of dramatic morphological change, ready-made coping mechanisms waiting there for future challenges.

Here is a discovery to lift the known from the ground and set it afloat before the eye.

·　·　·

When I first learned about evolution, I saw the process as a long, straight line hooked to bare rock in some dark early sea, slowly twisting in unbroken continuity through the ages. I saw new, more complex species bursting off the line at intervals, propelled by random "point" mutations of individual genes. From simple forms, complexity took shape with the slight mutations of genes.

The view was pleasing. It's in our nature to purify the chaotic world with lines and scales, with pyramids, ladders, and trees, all life neatly stepped, runged, branched. We like to search for simple cause and effect, for absolute order, to imagine life forms ratcheting forward from little to big — sessile and unseeing to mobilized and sighted, silent and unfeeling to anxious and talkative — all with minimal overlapping, minimal slipping back, as if life were an orderly continuum, like the growing of a crystal, as if evolution danced along a straight path like electricity. But the close kinship of genes across radically different creatures and the hopping of genes from one species to another unravels these notions, informs us that the beloved crowd of life forms on our planet exist not as discrete steps on a continuous stairway to our own exalted being but as a thick, fabulous tapestry.

This is good news. If the genes and cell mechanisms underlying growth, differentiation, reproduction, development, and adaptation are common to all life, then they can be studied conveniently, sometimes in one organism, sometimes in another. If life does indeed share so much ground, then we can fly to likeness as we do to beauty, as refuge from the terrors of being finite, specific, alone.

I find that keeping in mind both the multiplicity of life and the high levels of oneness — both the wavelengths of the rainbow and its colored arc — firms my thinking. Likewise, considering at once the power and persistence of individual genes and their subtle, elaborate linkage with the rest of life is good for pondering other instances of the many in the one — family, society, world.

Difficult as it is to see the beginnings of things, it is even harder to see the ends. The word "evolution" has come to describe all sorts of progressions from simple origins to more sophisticated ends. But this word, like the word "gene," is layered like an onion with older meanings that still affect its significance. "Evolution" goes back to the Latin *evolvere*, to unfold or disclose. Cicero used it to mean the unrolling and reading of a scroll. Perhaps we should return along this etymological route, to think of evolution not as a ladder leading to our house, but as an unfolding scroll shot through with mystery. The narrative is no neat line; it is a long, supple thread, at once loose enough to chance the new, yet tight enough to bind us to all the family that ever was.

SOURCES

I consulted many books and articles in order to piece together the story of the molecular family resemblances that unite life. Because this book is intended as a popular account rather than a scholarly study, I decided against using footnotes or mentioning in the text the names of all of the scientists whose work I have read and relied on. I owe them all my deep gratitude. The notes that follow are intended to direct the interested reader to the scientific papers I refer to in the text and to sources for further reading.

1. GENEALOGY

PAGE

4–5   I used Donald Frame's excellent translation of Michel de Montaigne's essay "Of the Resemblance of Children to Fathers" in *The Complete Essays of Montaigne* (Stanford University Press, 1998).

6–8   I refer to Victor McKusick, *Mendelian Inheritance in Man,* 12th ed. (Johns Hopkins University Press, 1998). The on-line version, known as OMIM (for Online Mendelian Inheritance in Man), can be found at www.ncbi.nlm.gov/omim/.

9–11   Among the works I consulted on the history of biological classification and early insights on the kinship of living things are Lisbet Koerner, *Linnaeus: Nature and Nation* (Harvard University Press, 1999); Stephen Jay Gould, *Ever Since Darwin* (Norton, 1977), and "More Light on Leaves," *Natural History,* February 1991, p. 16; Peter F. Stevens, *The Development of Biological Systematics* (Columbia University Press, 1994), and *Plants and Animals, Form and Relationship,* exhibit catalogue (Harvard University Herbaria, 1998); Tore Frängsmyr, ed., *Linnaeus: The Man and His Work* (Science History Publications, 1994); and François Jacob, *The Logic of Life* (Princeton University Press, 1973).

12–13   Discussion of the discovery of gene families and the new understanding of the tree of life can be found in Steven Henikoff et al., "Gene Families: The Taxonomy of Protein Paralogs and Chimeras," *Science* 278 (1997): 609; Rebecca Clayton et al., "The First Genome from the Third Domain of Life," *Nature* 387 (1997): 459; Jonathan Hodgkin, "A View of Mount *Drosophila*," *Nature* 404 (2000): 442; Russell Doolittle, "Microbial Genomes Opened Up," *Nature* 392 (1998): 339; W. Ford Doolittle, "Phylogenetic Classification and the Universal Tree," *Science* 284 (1999): 2124, and "Uprooting the Tree of Life," *Scientific American,* February 2000, p. 90; Carl Woese et al., "Towards a Natural System of Organisms: Proposal for the Domains Archaea, Bacteria, and Eukarya," *Proceedings of the National Academy of Sciences, USA* 87 (1990): 4576; G. Olsen and C. Woese, "Lessons from an Archaeal Genome: What Are We Learning from *Methanococcus jannaschii?*" *Trends in Genetics* 12 (1996): 377; and Mark Pagel, "Inferring the Historical Patterns of Biological Evolution," *Nature* 401 (1999): 877.

2. THE LONGEST THREAD

The literature on DNA is immense. For historical perspective, I relied on Horace Freeland Judson, *The Eighth Day of Creation* (Simon and Schuster, 1979); and François Jacob, *Logic of Life,* and *Of Flies, Mice, and Men* (Harvard University Press, 1998). Particularly valuable sources on the workings of DNA are Bruce Alberts et al., *The Molecular Biology of the Cell,* 3rd ed. (Garland Publishing, 1994); and Robert Pollack, *Signs of Life: The Language and Meanings of DNA* (Houghton Mifflin, 1994).

15–16   For the discovery of DNA's structure, I referred to James Watson and Francis Crick, "Molecular Structure of Nucleic Acids: A Structure for Deoxyribose Nucleic Acid," *Nature* 171 (1953): 737, and "Genetical Implications of the Structure of Deoxyribonucleic Acid," *Nature* 171 (1953): 964; and Joshua Lederberg, "What the Double Helix (1953) Has Meant for Basic Biomedical Science," *Journal of the American Medical Association* 269, no. 15 (1993): 1981.

16–20   Good sources on DNA duplication and repair include Miroslav Radman and Robert Wagner, "The High Fidelity of DNA Duplication," *Scientific American*, August 1988: 40; and Errol C. Friedberg, *Correcting the Blueprint of Life* (Cold Spring Harbor Laboratory Press, 1997). The two papers referred to in the text on the benefits of rapid mutation are Martin Wikelski and Corrina Thom, "Marine Iguanas Shrink to Survive El Niño," *Nature* 403 (2000): 37; and E. Richard Moxon and Christopher F. Higgins, "*E. coli* Genome Sequence: A Blueprint for Life," *Nature* 389 (1997): 20.

20–21   I refer to the Borges essay "The Analytical Language of John Wilkins," in *Other Inquisitions 1937 1952*, trans. Ruth L. C. Simms (Simon and Schuster, 1964).

22–23   For a discussion of hemoglobin evolution, see Vernon Ingram, *The Hemoglobins in Genetics and Evolution* (Columbia University Press, 1963); and R. E. Dickerson and I. Geis, *Hemoglobin: Structure, Function, Evolution, and Pathology* (Benjamin-Cummings, 1983). I quote from François Jacob, *The Possible and the Actual* (University of Washington, 1982).

24–28   Among the many interesting articles on junk DNA and jumping genes are Walter Gilbert, "On the Ancient Nature of Introns," *Gene* 135 (1993): 137; Petter Portin, "The Concept of the Gene: Short History and Present Status," *Quarterly Review of Biology* 68, no. 2 (1993): 173; and Nina Fedoroff and David Botstein, eds., *The Dynamic Genome: Barbara McClintock's Ideas in the Century of Genetics* (Cold Spring Harbor Laboratory Press, 1992). For more on McClintock's work, see Barbara McClintock, "Chromosome Arrangement and Genic Expression," *Cold Spring Harbor Symposium on Quantitative Biology* 16 (1951): 13; and Evelyn Fox Keller's fine biography of McClintock, *A Feeling for the Organism* (W. H. Freeman, 1983).

3. CHANCE IN THE HOUSE OF FATE

29   Accounts of Williams syndrome were taken from S. J. Paterson et al., "Cognitive Modularity and Genetic Disorders," *Science*

286 (1999): 2355; and Annette Karmiloff-Smith, "Development Itself Is the Key to Understanding Developmental Disorders," *Trends in Cognitive Sciences* 2, no. 10 (1998): 389.

30–39   I found excellent descriptions of proteins, protein structure, and protein folding in *Finding the Critical Shapes* (Howard Hughes Medical Institute, 1990); George D. Rose, "No Assembly Required," *The Sciences*, January/February 1996, p. 26; F. Ulrich Hartl, "Molecular Chaperones in Cellular Protein Folding," *Nature* 381 (1996): 571; and David Eisenberg, "How Chaperones Protect Virgin Proteins," *Science* 285 (1999): 1021.

31–32   I refer to Geerat Vermeij, *A Natural History of Shells* (Princeton University Press, 1993).

33–34   Engaging descriptions of Leeuwenhoek and the history of microscopy can be found in Catherine Wilson, *The Invisible World: Early Modern Philosophy and the Invention of the Microscope* (Princeton University Press, 1995); Brian J. Ford, *A Single Lens* (Harper and Row, 1985); Clifford Dobell, *Antony van Leeuwenhoek and His "Little Animals"* (Dover, 1960); and A. Schierbeek, *Measuring the Invisible World: The Life and Works of Antoni van Leeuwenhoek* (Abelard-Schuman, 1959).

36–39   For information on the evolution of proteins and protein motifs, see Robert L. Dorit, "How Big Is the Universe of Exons?" *Science* 250 (1990): 1377; Russell F. Doolittle, "Microbial Genomes Opened Up," *Nature* 392 (1998): 339; "Determining Divergence Times of the Major Kingdoms of Living Organisms with a Protein Clock," *Science* 271 (1996): 70; "The Origins and Evolution of Eukaryotic Proteins," *Philosophical Transactions of the Royal Society of London B* 349 (1995): 235; and "The Multiplicity of Domains in Proteins," *Annual Review of Biochemistry* 64 (1995): 287.

4. SEEDS OF INHERITANCE

43–44   A good discussion of cloning can be found in Anne McLaren, "Cloning: Pathways to a Pluripotent Future," *Science* 288 (2000): 1775.

44–46   For a detailed and fascinating account of early beliefs about eggs, sperm, and embryology, see Clara Pinto-Correia, *The*

Ovary of Eve: Eggs and Sperm and Preformation (University of Chicago Press, 1997). See also Jacob, *Logic of Life.*

46   I refer to Hubert Schwabl's paper "Yolk Is a Source of Maternal Testosterone for Developing Birds," *Proceedings of the National Academy of Sciences, USA* 90 (1993): 11446.

46–47   Quotations from *De Secretis Mulierum* come from Helen Rodnite Lemay, *Women's Secrets: A Translation of Pseudo-Albertus Magnus's De Secretis Mulierum with Commentaries* (State University of New York Press, 1992).

47–49   For information on sperm, sperm competition, and sexual selection, see Tim Birkhead, *Sperm Competition and Sexual Selection* (Academic Press, 1998), and "Distinguished Sperm in Competition," *Nature* 400 (1999):46; and William G. Eberhard, *Female Control: Sexual Selection by Cryptic Female Choice* (Princeton University Press, 1996).

48–49   Theory on the evolution of sex cells is discussed in G. A. Parker et al., "The Origin and Evolution of Gametic Dimorphism and the Male-Female Phenomenon," *Journal of Theoretical Biology* 26 (1972): 529.

50–51   I found material on the chemical dialogue between egg and sperm in Kathleen R. Foltz et al., "Sea Urchin Egg Receptor for Sperm," *Science* 259 (1993): 1421, and "Gamete Recognition and Egg Activation in Sea Urchins," *American Zoologist* 35 (1995): 381; Dina Ralt et al., "Sperm Attraction to a Follicular Factor(s) Correlates with Human Egg Fertilizability," *Proceedings of the National Academy of Sciences USA* 88 (1991): 2840; and Anat Cohen-Dayag et al., "Sperm Capacitation in Humans Is Transient and Correlates with Chemotactic Responsiveness to Follicular Factors," *Proceedings of the National Academy of Sciences, USA* 92 (1995): 11039.

51   Research on proteins involved in gamete recognition is described in Victor D. Vacquier, "Evolution of Gamete Recognition Proteins," *Science* 281 (1998): 1995; Edward Metz and Stephen Palumbi, "Positive Selection and Sequence Rearrangements Generate Extensive Polymorphism in the Gamete Recognition Protein Bindin," *Molecular Biology and Evolution* 13, no. 2 (1996): 392; William J. Snell and Judith M. White, "The Mole-

cules of Mammalian Fertilization," *Cell* 85 (1996): 629; and Willie Swanson and Victor Vacquier, "Concerted Evolution in an Egg Receptor for a Rapidly Evolving Abalone Sperm Protein," *Science* 281 (1998): 710.

52 The quotation from Laurence Hurst comes from his paper "Selfish Genetic Elements and Their Role in Evolution: The Evolution of Sex and Some of What That Entails," *Philosophical Transactions of the Royal Society of London B* 349 (1995): 321.

52 I refer to Adam Eyre-Walker's study on the rate of mutation in the human genome, "High Genomic Deleterious Mutation Rates in Hominids," *Nature* 397 (1999): 344.

53 For an excellent account of David Haig's theory, see his article "Genetic Conflicts in Human Pregnancy," *Quarterly Review of Biology* 68 (1993): 495.

5. CELLE FANTASTYK

54–57 I found a detailed and compelling history of cell biology in Henry Harris, *The Birth of the Cell* (Yale University Press, 1999); and also in Lewis Wolpert, "Evolution of the Cell Theory," *Philosophical Transactions of the Royal Society of London B* 349 (1995): 227.

57–58 Engaging descriptions of cephalopods are found in Martin J. Wells, *Octopus: Physiology and Behavior of an Advanced Invertebrate* (Chapman and Hall, 1978).

59–60 An excellent source for general information about cell biology is Bruce Alberts, *Essential Cell Biology* (Garland Publishing, 1998). Similarities in cell cycle proteins are reported in Bruce A. Edgar and Christian F. Lehner, "Developmental Control of Cell Cycle Regulators: A Fly's Perspective," *Science* 274 (1996): 1646.

60–61 Among the many fine references on the evolution of the complex cell are Betsey Dexter Dyer, *Tracing the History of Eukaryotic Cells* (Columbia University Press, 1994); Christian de Duve, "The Birth of Complex Cells," *Scientific American,* April 1996, p. 50; Jan Sapp, *Evolution by Association: A History of Symbiosis* (Oxford University Press, 1994); and Lynn Margulis, *Symbiosis in Cell Evolution* (W. H. Freeman, 1993).

62–64   Research on cell fate and cell lineage is reported in *From Egg to Adult* (Howard Hughes Medical Institute, 1992); and J. Sulston and H. R. Horvitz, "Postembryonic Cell Lineages of the Nematode *Caenorhabditis elegans*," *Developmental Biology* 56 (1977): 110.

6. GENERATION

66–67   The discovery of *Symbion pandora* is described in Simon Conway Morris, "A New Phylum from the Lobster's Lips," *Nature* 378 (1995): 661.

68   My reference to the link between conceptual categories in the brain and their lexical retrieval comes from Hanna Damasio, "A Neural Basis for Lexical Retrieval," *Nature* 380 (1996): 499.

68–69   Accounts of the Burgess Shale and the Cambrian explosion are found in Stephen Jay Gould's brilliant and entertaining book *Wonderful Life* (Norton, 1989) and in Jeffrey S. Levinton, "The Big Bang of Animal Evolution," *Scientific American*, November 1992, p. 84; Andrew Knoll, "Breathing Room for Early Animals," *Nature* 382 (1996): 111; and Graham Logan, "Terminal Proterozoic Reorganization of Biogeochemical Cycles," *Nature* 357 (1995): 238.

69–70   Among the many references on symmetry in the animal world are Randy Thornhill, "The Allure of Symmetry," *Natural History*, September 1993, p. 30; Anders Pape Møller, "Female Swallow Preference for Symmetrical Male Sexual Ornaments," *Nature* 357 (1992): 238; Natalic Angier, "Why Birds and Bees, Too, Like Good Looks," *New York Times*, February 8, 1994, p. C1; and Steve Blinkhorn, "Symmetry as Destiny — Taking a Balanced View of IQ," *Nature* 387 (1997): 849.

71–72   For the history of early beliefs about embryology and development, see Pinto-Correia, *Ovary of Eve*.

72–76   The story of Hox genes and other developmental genes is told in Lewis Wolpert, *Principles of Development* (Oxford University Press, 1998); Rudolf A. Raff, *The Shape of Life: Genes, Development, and the Evolution of Animal Form* (University of Chicago Press, 1996); John Gerhart and Marc Kirschner, *Cells, Embryos and Evolution* (Blackwell Science, 1997); Matthew P. Scott, "Structural Relationships among Genes That Control Development: Sequence Homology Between the Antennapedia, Ultrabithorax, and Fushi Tarazu Loci of *Drosophila*," *Proceedings of the National Academy of Sciences, USA* 81 (1984): 4115, and "Hox Proteins Reach Out Round DNA," *Nature* 397 (1999): 649; William McGinnis, "The Molecular Architects of Body Design," *Scientific American,* February 1994, p. 58; Walter J. Gehring, *Master Control Genes in Development and Evolution: The Homeobox Story* (Yale University Press, 1999); Eddy De Robertis et al., "Homeobox Genes and the Vertebrate Body Plan," *Scientific American,* July 1990, p. 46; Jim Smith, "How To Tell a Cell Where It Is," *Nature* 381 (1996): 367; Peter W. H. Holland, "The Future of Evolutionary Developmental Biology," *Nature* 402 (suppl.) (1999): C41; Denis Duboule, "A Hox by Any Other Name," *Nature* 403 (2000): 607; and Mark Martindale and Matthew Kourakis, "Hox Clusters: Size Doesn't Matter," *Nature* 399 (1999): 730.

75   I found good information on the genes and proteins involved in determining the body's asymmetry in Aimee K. Ryan et al., "Pitx2 Determines Left-Right Asymmetry of Internal Organs in Vertebrates," *Nature* 394 (1998): 545; Juan Carlos Izpisua Belmonte, "How the Body Tells Left from Right," *Scientific American,* June 1999, p. 46; and Shigenori Nonaka et al., "Randomization of Left-Right Asymmetry Due to Loss of Nodal Cilia Generating Leftward Flow of Extraembryonic Fluid in Mice Lacking KIF3B Motor Protein," *Cell* 95 (1998): 829.

75–76   Discussion of the evolution of Hox genes is found in Michael Akam, "Hox Genes and the Evolution of Diverse Body Plans," *Philosophical Transactions of the Royal Society of London B* 349 (1995): 313; and Frank H. Ruddle et al., "Evolution of *Hox* Genes," *Annual Review of Genetics* 28 (1994): 423.

76   For more information on the exciting discoveries about the birth of new cells in adult brains, see Gerd Kempermann and Fred H. Gage, "New Nerve Cells for the Adult Brain," *Scientific American,* May 1999, p. 48.

7. NEW TRICKS

77–78   For the work of Christiane Nüsslein-Volhard, I drew from the following papers: Christiane Nüsslein-Volhard and Eric Wieschaus, "Mutations Affecting Segment Number and Polarity in *Drosophila*," *Nature* 287 (1980): 795; Christiane Nüsslein-Volhard, "Of Flies and Fishes," *Science* 266 (1994): 572; and Adam Felsenfeld, "Defining the Boundaries of Zebrafish Developmental Genetics," *Nature Genetics* 14 (1996): 258.

78–81   Compelling accounts of the similarities between these body-patterning gene pathways can be found in Stephen Jay Gould, "Of Mice and Mosquitoes," *Natural History*, July 1991, p. 12; Eddy M. De Robertis and Yoshiki Sasai, "A Common Plan for Dorsoventral Patterning in Bilateria," *Nature* 380 (1996): 37; and Gary Ruvkun and Oliver Hobert, "The Taxonomy of Developmental Control in *Caenorhabditis elegans*," *Science* 282 (1998): 2033.

79–81   For an engaging account of the debate, see Toby A. Appel, *The Cuvier-Geoffroy Debate: French Biology in the Decades Before Darwin* (Oxford University Press, 1987). See also Stephen J. Gould, "As the Worm Turns," *Natural History*, February 1997, p. 24.

81–83   Several articles illuminate the work of Sean Carroll and others studying evolution and development, among them Sean Carroll, "Homeotic Genes and the Evolution of Arthropods and Chordates," *Nature* 376 (1995): 479; H. F. Nijhout, "Focus on Butterfly Eyespot Development," *Nature* 384 (1996): 209; Paul Brakefield et al., "Development, Plasticity and Evolution of Butterfly Eyespot Patterns," *Nature* 384 (1996): 236; Neil Shubin et al., "Fossils, Genes and the Evolution of Animal Limbs," *Nature* 388 (1997): 639; Stephen Gaunt, "Chick Limbs, Fly Wings and Homology at the Fringe," *Nature* 386 (1997): 324; Robert Riddle and Clifford Tabin, "How Limbs Develop," *Scientific American*, February 1999, p. 74; Andrew Knoll and Sean Carroll, "Early Animal Evolution: Emerging View from Comparative Biology and Geology," *Science* 284 (1999): 2129; and Douglas Erwin et al., "The Origin of Animal Body Plans," *American Scientist*, March-April 1997, p. 126.

8. SIGHTING LIFE

85–90 A good essay on the basic biology of the animal eye is Michael Land, "Optics of the Eyes of the Animal Kingdom," in John Cronly-Dillon, ed., *Evolution of the Eye and Visual System* (CRC Press, 1991), p. 118.

86 For more on the ancients' understanding of vision, see David C. Lindberg, *Theories of Vision from Al-Kindi to Kepler* (University of Chicago Press, 1976).

87–89 Discussions of the evolution of the eye are found in Richard Dawkins's engaging and informative *Climbing Mount Improbable* (Norton, 1996); and in Michael Land and Russell Fernald, "The Evolution of Eyes," *Annual Reviews of Neuroscience* 15 (1992): 1. The study by Dan-E. Nilsson and Susanne Pelger, "A Pessimistic Estimate of the Time Required for an Eye to Evolve," was published in *Proceedings of the Royal Society of London B* 256 (1994): 53.

89–90 I found Richard Gregory's account of his search for *Copilia quadrata* in "Origins of Eyes — with Speculations on Scanning Eyes," in *Evolution of the Eye and Visual System,* ed. John Cronly-Dillon (CRC Press, 1991), and in "See Naples and Live: The Scanning Eye of *Copilia,*" in Richard Gregory, *Odd Perceptions* (Methuen and Routledge, 1986).

90–91 My information on the common components of eyes comes from Russell Doolittle, "Lens Proteins: More Molecular Opportunism," *Nature* 336 (1988): 18. The passage on the evolution of rhodopsins comes from Shozo Yokoyama and T. Tada, "Adaptive Evolution of the African and Indonesian Coelacanths to Deep-Sea Environments," reported in Elizabeth Pennisi, "Gaining New Insight into the Molecular Basis of Evolution," *Science* 285 (1999): 654; and Keith Crandall and David Hillis, "Rhodopsin Evolution in the Dark," *Nature* 387 (1997): 667.

92–93 The unfolding story of the *eyeless* gene is told in several articles, including Stephen Jay Gould, "Common Pathways of Illumination," *Natural History,* December 1994, p. 10; Dan-E. Nilsson, "Eye Ancestry: Old Genes for New Eyes," *Current Biology* 6, no. 1 (1996): 39; R. Quiring et al., "Homology of the *Eyeless* Gene of

Drosophila to the *Small Eye* Gene in Mice and *Aniridia* in Humans," *Science* 265 (1994): 785; Georg Halder et al., "Induction of Ectopic Eyes by Targeted Expression of the Eyeless Gene in *Drosophila*," *Science* 267 (1995): 1788, and "New Perspectives on Eye Evolution," *Current Opinion in Genetics and Development* 5 (1995): 602; and P. H. Mathers et al., "The *Rx* Homeobox Gene Is Essential for Vertebrate Eye Development," *Nature* 387 (1997): 603.

94   My information on the theory that eyes may have evolved from light-sensitive skin cells comes from Heinz Arnhelter, "Eyes Viewed from the Skin," *Nature* 391 (1998): 632. The speculation on iridescent Cambrian creatures and their effect on vision comes from Andrew Parker, "Colour in Burgess Shale Animals and the Effect of Light on Evolution in the Cambrian," *Proceedings: Biological Sciences of the Royal Society,* June 7, 1998, p. 967.

9. THE CATERPILLAR FACULTY

95–98   Engaging accounts of the life and work of Maria Sibylla Merian can be found in Natalie Zemon Davis, *Women on the Margins: Three Seventeenth-Century Lives* (Harvard University Press, 1995); Elisabeth Rücker and William Stearn, *Maria Sibylla Merian in Surinam* (Pion, 1982); and Sharon D. Valiant, "Questioning the Caterpillar," *Natural History,* December 1992, p. 47. Francesco Redi's work is described in his *Experiments on the Generation of Insects,* trans. Mab Bigelow from 1688 Italian edition (Open Court Publishing, 1909).

98–104 I gleaned information about the discovery and workings of cell death from many articles, among them Peter Clarke and Stephanie Clarke, "Nineteenth-Century Research on Naturally Occurring Cell Death and Related Phenomena," *Anatomical Embryology* 193 (1996): 81; Richard Lockshin et al., *When Cells Die* (Wiley-Liss, 1998); A. Glucksmann, "Cell Deaths in Normal Vertebrate Ontogeny," *Biological Reviews* 26 (1951): 59; Michael Jacobson et al., "Programmed Cell Death in Animal Development," *Cell* 88 (1997): 347; Martin Raff, "Cell Suicide for Beginners," *Nature* 396 (1998): 119; Martin Raff et al., "Social Controls on Cell Survival and Cell Death," *Nature* 356 (1992): 397, and "Programmed Cell Death and the Control of Cell Survival: Lessons from the Nervous System," *Science* 262 (1993): 695; Richard Duke et al., "Cell Suicide in Health and Disease," *Scientific American*, December 1996, p. 80; Gerard Evan and Trevor Littlewood, "A Matter of Life and Cell Death," *Science* 281 (1998): 1317; Ronald Ellis et al., "Mechanisms and Functions of Cell Death," *Annual Review of Cell Biology* 7 (1991): 663; and Seamus Martin and Douglas Green, "Protease Activation During Apoptosis: Death by a Thousand Cuts," *Cell* 82 (1995): 349.

100–104   Descriptions of the work of Bob Horvitz and others on the genetics of cell death are found in H. R. Horvitz, "Programmed Cell Death in Nematode Development," *NeuroScience Commentaries* 1 (1982): 56; H. R. Horvitz and H. Ellis, "Genetic Control of Programmed Cell Death in the Nematode *C. elegans*," *Cell* 44 (1986): 817 ; J. Yuan and H. R. Horvitz, "The *Caenorhabditis elegans* Genes *ced-3* and *ced-4* Act Autonomously to Cause Programmed Cell Death," *Developmental Biology* 138 (1990): 33; M. Hengartner et al., "*Caenorhabditis elegans* Gene *ced-9* Protects Cells from Programmed Cell Death," *Nature* 356 (1992): 494; M. Hengartner and H. R. Horvitz, "*C. elegans* Cell Survival Gene *ced-9* Encodes a Functional Homolog of the Mammalian Proto-oncogene *bcl-2*," *Cell* 76 (1994): 665; and David Vaux et al., "Prevention of Programmed Cell Death in *Caenorhabditis elegans* by Human *bcl-2*," *Science* 258 (1992): 1955.

101   For theories on the evolutionary roots of cell death, see Jean Ameisen, "The Origin of Programmed Cell Death," *Science* 272 (1996): 1278. Alcohol-induced apoptosis in the fetus is the subject of Chrysanthy Ikonomidou, "Ethanol-Induced Apoptotic Neurodegeneration and Fetal Alcohol Syndrome," *Science* 287 (2000): 1056. Studies on infant maternal deprivation by Mark A. Smith of DuPont Merck Research Laboratory in Wilmington were reported in *Science News* 152 (1997): 298.

102   Information on the role of mitochondria in cell death can be found in Catherine Brenner and Guido Kroemer, "Mitochondria — the Death Signal Integrators," *Science* 289 (2000): 1150.

103   The p53 gene and cell death are discussed in D. P. Lane, "p53, Guardian of the Genome," *Nature* 358 (1992): 15; and L. J. Ko and C. Prives, "p53: Puzzle and Paradigm," *Genes and Development* 10 (1996): 1054.

10. SEXING LIFE

105   For information on Hox genes in the snake, see Martin Cohn, "Developmental Basis of Limblessness and Axial Patterning in Snakes," *Nature* 399 (1999): 474.

106–7   Methods of sexing animals are described in Kate Lessels and Christa Mateman, "Molecular Sexing of Birds," *Nature* 383 (1996): 761; Tim Guilford and Paul Harvey, "The Purple Patch," *Nature* 392 (1998): 867; and Darryl Gwynne, "Genitally Does It," *Nature* 393 (1998): 734.

107–8   I refer to Peter Fryer's *Mrs. Grundy: Studies in English Prudery* (London House and Maxwell, 1963).

109   The quote is from Pollack, *Signs of Life.*

110–11   Historical beliefs on the origins of gender are described in Pinto-Correia, *Ovary of Eve.*

111–15   I found excellent accounts of sex determination mechanisms and the evolution of sex in David Crews, "Animal Sexuality," *Scientific American,* January 1994, p. 108; John Maynard Smith, *The Evolution of Sex* (Cambridge University Press, 1978); Lisa Ryner and Amanda Swain, "Sex in the '90s," *Cell* 81 (1995): 483; R. H. F. Hunter, *Sex Determination, Differentiation and Inter-sexuality in Placental Mammals* (Cambridge University Press, 1995); Bruce Lahn and David Page, "Four Evolutionary Strata on the Human X Chromosome," *Science* 286 (1999): 964; Michael Hammer, "A Recent Common Ancestry for Human Y Chromosomes," *Nature* 378 (1995): 376; Ignacio Marin and Bruce Baker, "The Evolutionary Dynamics of Sex Determination," *Science* 281 (1998): 1990; Jonathan Hodgkin, "Genetic Sex Determination Mechanisms and Evolution," *BioEssays* 14, no. 4 (1992): 253, and "Sex Determination Compared in *Drosophila* and *Caenorhabditis*," *Nature* 344 (1990): 721.

113   For evidence of the genetic and hormonal complexity of sex-

determination mechanisms in mammals, see Seppio Vainio, "Female Development in Mammals Is Regulated by Wnt-4 Signalling," *Nature* 397 (1999): 405; and Rex Hess et al., "A Role for Oestrogens in the Male Reproductive System," *Nature* 390 (1997): 509.

114   I refer to Craig Smith et al., "Conservation of a Sex-Determining Gene," *Nature* 402 (1999): 601; and Christopher Raymond et al., "Evidence for Evolutionary Conservation of Sex-Determining Genes," *Nature* 391 (1998): 691.

115–16   This and other theories of sex and conflct are outlined in Laurence Hurst and William Hamilton, "Cytoplasmic Fusion and the Nature of Sexes," *Proceedings of the Royal Society of London B* 247 (1992): 189; Linda Partridge and Laurence Hurst, "Sex and Conflict," *Science* 281 (1998): 2003; and Laurence Hurst et al., "Genetic Conflicts," *Quarterly Review of Biology* 71, no. 3 (1996): 317.

116   For information on bdelloid rotifers, see David Mark Welch and Matthew Meselson, "Evidence for the Evolution of Bdelloid Rotifers Without Sexual Reproduction or Genetic Exchange," *Science* 288 (2000): 1211.

116–17   I found material on the debate about the evolution of sex in N. H. Barton and B. Charlesworth, "Why Sex and Recombination?" *Science* 281 (1998): 1986; William Hamilton et al., "Sexual Reproduction as an Adaptation to Resist Parasites," *Proceedings of the National Academy of Sciences, USA* 87 (1990): 3566; and David Crews, "The Evolutionary Antecedents to Love," *Psychoneuroendocrinology* 23, no. 8 (1998): 751.

11. THE MOTH BENEATH THE SKIN

Of the many references on smell, I found especially informative D. Michael Stoddart, *The Scented Ape: The Biology and Culture of Human Odour* (Cambridge University Press, 1991); Steve Van Toller and George Dodd, *Perfumer: The Psychology and Biology of Fragrance* (Routledge, Chapman and Hall, 1988); William Agosta, *Chemical Communication: The Language of Pheromones* (Scientific American Library, 1992); Piet Vroon, *Smell* (Farrar, Straus, and Giroux, 1994); and Lyall Watson, *Jacobson's Organ: And the Remarkable Nature of Smell* (Norton, 2000).

121   I refer to a study by Peter Hepper, "The Amniotic Fluid: An Important Priming Role in Kin Recognition," *Animal Behavior* 35 (1987): 1343.

121–24   I drew information on the smell skills of various animals from Jelle Atema, "Chemical Signals in the Marine Environment," *Proceedings of the National Academy of Sciences, USA* 92 (1995): 62; H. Kalmus, "The Discrimination by the Nose of the Dog of Individual Human Odours and in Particular of the Odours of Twins," *Animal Behavior* 3 (1955): 25; and Agosta, *Chemical Communication.*

123   For Fabre's compelling version of the story, see his *Life of the Caterpillar* (Dodd, Mead, 1916).

124–26   I found discussions of historical attitudes toward smell and odors in William S. Cain, "History of Research on Smell," in *Handbook of Perception*, vol. 6A (Academic Press, 1978); in Stoddart, *Scented Ape;* Annick LeGuérer, *Scent* (Kodansha, 1994); Van Toller and Dodd, *Perfumer;* and Vroon, *Smell.*

128   The quote from the anosmic man comes from Oliver Sacks, *The Man Who Mistook His Wife for a Hat* (Harper and Row, 1985), p. 156

126–29   A fine introduction to the anatomy and molecular biology of smell is in *Seeing, Hearing, and Smelling the World* (Howard Hughes Medical Institute, 1995). Also useful is John Hildebrand, "Analysis of Chemical Signals by Nervous Systems," *Proceedings of the National Academy of Sciences, USA* 92 (1995): 67; and Kensaku Mori et al., "The Olfactory Bulb: Coding and Processing of Odor Molecule Information," *Science* 286 (1999): 703.

128–29   For descriptions of smell receptors and the story of their discovery, see Linda Bartoshuk and Gary Beauchamp, "Chemical Senses," *Annual Review of Psychology* 1994: 419; Linda Buck and Richard Axel, "A Novel Multigene Family May Encode Odorant Receptors: A Molecular Basis for Odor Recognition," *Cell* 65 (1991): 175; Richard Axel, "The Molecular Logic of Smell," *Scientific American*, October 1995, p. 154; Linda Buck, "Unraveling Chemosensory Diversity," *Cell* 83 (1995): 349; B. Malnic et al., "Combinatorial Receptor Codes for Odors," *Cell* 96 (1999): 713; Peter Mombaerts, "Seven-Transmembrane Proteins as Odorant and Chemosensory Receptors," *Science* 286 (1999): 707; Maurine Linder and Alfred Gilman, "G Proteins," *Scientific American,* July 1992, p. 56; Haiqing Zhao et al., "Functional Expression of a Mammalian Odorant Receptor," *Science* 279 (1998): 237; and Gilles Laurent, "A Systems Perspective on Early Olfactory Coding," *Science* 286 (1999): 7223.

129   For information on smell receptors in flies and other invertebrates, I referred to Yitzhak Pilpel and Doron Lancet, "Good Reception in Fruitfly Antennae," *Nature* 398 (1999): 285; and Jürgen Krieger and Heinz Breer, "Olfactory Reception in Invertebrates," *Science* 286 (1999): 720.

129–30   The comparative study of smell genes in primates and rodents is S. Rouquier et al., "The Olfactory Receptor Gene Repertoire in Primates and Mouse: Evidence for the Reduction of the Functional Fraction in Primates," *Proceedings of the National Academy of Sciences, USA* 97 (2000): 2870.

130   I refer to Oliver Sacks, *The Man Who Mistook His Wife for a Hat* (Harper and Row, 1985).

131   I found the discussion of cases of hyperosmia in Harry Wiener, "External Chemical Messengers," *New York State Journal of Medicine,* 15 December 1966, p. 3153; and A. A. Brill, "The Sense of Smell in the Neuroses and Psychoses," *Psychoanalytic Quarterly* 1 (1932): 7.

130–33   For detailed descriptions of pheromones in the animal world, see Agosta, *Chemical Communication;* and Edward O. Wilson, "Pheromones," *Scientific American,* May 1963, p. 100.

133   Studies on the possible role of pheromones and pheromone receptors in human biology include Martha McClintock, "Menstrual Synchrony and Suppression," *Nature* 229 (1971): 244; Leonard Weller and Aron Weller, "Human Menstrual Synchrony: A Critical Assessment," *NeuroScience and Biobehavioral Reviews* 17 (1993): 427; Trese Leinders-Zufall et al., "Ultrasensitive Pheromone Detection by Mammalian Vomeronasal Neurons," *Nature* 405 (2000): 792; Kathleen Stern and Martha McClintock, "Regulation of Ovulation by Human Pheromones," *Nature* 392 (1998): 177; Aron Weller, "Communication Through Body Odor," *Nature* 392 (1998): 16; Catherine Dulac and Richard

Axel, "A Novel Family of Genes Encoding Putative Pheromone Receptors in Mammals," *Cell* 83 (1995): 195; Hiroaki Matsunami and Linda Buck, "A Multigene Family Encoding a Diverse Array of Putative Pheromone Receptors in Mammals," *Cell* 90 (1997): 775; David Moran, Bruce Jafek, and J. Carter Rowley III, "The Vomeronasal (Jacobson's) Organ in Man: Ultrastructure and Frequency of Occurrence," *Journal of Steroid Biochemistry and Molecular Biology* 39, no. 4B (1991): 545; Eric Keverne, "The Vomeronasal Organ," *Science* 286 (1999): 716; and Ivan Rodriguez et al., "A Putative Pheromone Receptor Gene Expressed in Human Olfactory Mucosa," *Nature Genetics* 26, no. 1 (2000): 18.

12. TWO BROTHERS, EIGHT COUSINS

134   Evidence of an infant's ability to recognize its mother is presented in A. Macfarlane, "Olfaction in the Development of Social Preferences in the Human Neonate," *The Human Neonate in Parent-Infant Interaction*, CIBA Foundation Symposium 33 (1975): 103; and Jennifer Makin and Richard Porter, "Attractiveness of Lactating Females' Breast Odors to Neonates," *Child Development* 60 (1989): 803.

134–36   For fascinating descriptions of the abilities of plants and animals to recognize kin, see David Pfennig and Paul Sherman, "Kin Recognition," *Scientific American*, June 1995, p. 97; Peter Hepper, *Kin Recognition* (Cambridge University Press, 1991); Richard Grosberg and James Quinn, "The Genetic Control and Consequences of Kin Recognition by the Larvae of a Colonial Marine Invertebrate," *Nature* 322 (1986): 456; Gerard Arnold et al., "Kin Recognition in Honeybees," *Nature* 379 (1996): 498; David Pfennig et al., "Kin Recognition and Cannibalism in Spadefoot Toad Tadpoles," *Animal Behavior* 46 (1993): 87, and "Kin Recognition and Cannibalism in Polyphenic Salamanders," *Behavior Ecology* 5, no. 2 (1994): 225; David Pfennig, "Kinship Affects Morphogenesis in Cannibalistic Salamanders," *Nature* 362 (1993): 836; and Lisa Parrand and Frans B. M. de Waal, "Visual Kin Recognition in Chimpanzees," *Nature* 399 (1999): 64.

136–37   A detailed description of the biology of human odor can be found in Stoddart, *Scented Ape*.

137   For evidence of human kin recognition through smell, see R. H. Porter et al., "Recognition of Kin Through Characteristic Body Odors," *Chemical Senses* 11 (1986): 389; and R. H. Porter and J. D. Moore, "Human Kin Recognition by Olfactory Cues," *Physiology and Behavior* 27 (1981): 493. The quotation by Cabanis comes from "Rapport du Physique et du Moral" (1802) in *Oeuvres Complètes* (P.U.F.), 1956, cited by Annick LeGuérer in *Scent*.

137–38   I refer to the paper by Peter Hepper, "The Discrimination of Human Odour by the Dog," *Perception* 17, no. 4 (1988): 549.

138–39   In exploring the genetic basis of kin recognition, I referred to studies by K. Yamazaki et al., "Control of Mating Preferences in Mice by Genes in the Major Histocompatibility Complex," *Journal of Experimental Medicine* 144 (1976): 1324, and "Recognition Among Mice: Evidence from the Use of a Y-Maze Differentially Scented by Congenic Mice of Different Major Histocompatibility Types," *Journal of Experimental Medicine* 150 (1979): 755; Wayne Potts et al., "Mating Patterns in Seminatural Populations of Mice Influenced by MHC Genotype," *Nature* 352 (1991): 619; Avery Nelson Gilbert et al., "Olfactory Discrimination of Mouse Strains *(Mus musculus)* and Major Histocompatibility Types by Humans *(Homo sapiens),*" *Journal of Comparative Psychology* 100, no. 3 (1986): 262; Gary Beauchamp et al., "The Chemosensory Recognition of Genetic Individuality," *Scientific American,* July 1985, p. 86; Alan Grafen, "Of Mice and the MHC," *Nature* 360 (1992): 530; and Jerram Brown and Amy Eklund, "Kin Recognition and the Major Histocompatibility Complex: An Integrative Review," *American Naturalist* 143, no.3 (1994): 435.

139   My information on Hutterites comes from John Hostetler, *Hutterite Society* (Johns Hopkins University Press, 1974); and Victor Peters, *All Things Common: The Hutterian Way of Life* (University of Minnesota Press, 1965). For the Ober study, see Carole Ober et al., "HLA and Mate Choice in Humans," *American Journal of Human Genetics* 61 (1997): 497. But see also opposing evidence by Philip Hedrick and Francis Black, "HLA and

Mate Selection: No Evidence in South Amerindians," *American Journal of Human Genetics* 61 (1997): 505. A discussion of the two studies can be found in Gary Beauchamp and Kunio Yamazaki, "HLA and Mate Selection in Humans: Commentary," *American Journal of Human Genetics* 61 (1997): 494.

140   Inbreeding and its effects is the topic of an excellent anthology, Nancy Thornhill, ed., *The Natural History of Inbreeding and Outbreeding* (University of Chicago Press, 1993). I also found informative the following two papers on the effects of human inbreeding: Alan Bittles et al., "Reproductive Behavior and Health in Consanguineous Marriages," *Science* 252 (1991): 789; and Alan Bittles and James Neel, "The Costs of Human Inbreeding and Their Implications for Variations at the DNA Level," *Nature Genetics* 8 (1994): 117. The quotation from Dawkins comes from *The Selfish Gene* (Oxford University Press, 1989).

140–41   I refer to the study by Ilik Saccheri et al., "Inbreeding and Extinction in a Butterfly Metapopulation," *Nature* 392 (1998): 491.

141   Hamilton's ideas are set forth in W. D. Hamilton, "The Genetical Evolution of Social Behavior (I and II)," *Journal of Theoretical Biology* 7 (1964): 1, 17. I also found entertaining and informative Hamilton's essay "Inbreeding in Egypt and in This Book: A Childish Perspective," in Thornhill, *Natural History of Inbreeding and Outbreeding.*

141–42   I refer to Lewis Thomas's essay "Altruism," in *Late Night Thoughts on Listening to Mahler's Ninth Symphony* (Bantam, 1984).

13. A KIND OF REMEMBRANCE

144   My information on the molecules of memory comes from P. Goelet et al., "The Long and Short of Long-Term Memory," *Nature* 322 (1986): 419; J. C. P. Yin et al., "CREB as a Memory Modulator," *Cell* 81 (1995): 107; and Ya-Pin Tang et al., "Genetic Enhancement of Learning and Memory in Mice," *Nature* 40 (1999): 63.

145–54   I consulted several books and articles on the workings of the immune system, among them the excellent *Arousing the Fury of the Immune System* (Howard Hughes Medical Institute, 1998); Rafi Ahmed and David Gray, "Immunological Memory and Protective Immunity: Understanding Their Relation," *Science* 272 (1996): 54; Charles Janeway, Jr., "How the Immune System Recognizes Invaders," *Scientific American,* September 1993, p. 73; Arthur Silverstein, *A History of Immunology* (Academic Press, 1989); Louis du Pasquier, "Origin and Evolution of the Vertebrate Immune System," *APMIS* 100 (1992): 383; and Gregory Beck and Gail Habicht, "Immunity and the Invertebrates," *Scientific American,* November 1996, p. 60. Other sources are listed below under specific subject categories.

146   For information on natural antibiotics, see Michael Zasloff, "Magainins, a Class of Antimicrobial Peptides from *Xenopus* Skin," *Proceedings of the National Academy of Sciences, USA* 84 (1987): 5449; Barry Schonwetter et al., "Epithelial Antibiotics Induced at Sites of Inflammation, *Science* 287 (1995): 1645; J. Harder et al., "A Peptide Antibiotic from Human Skin," *Nature* 387 (1997): 861; Robert Hancock, "Peptide Antibiotics," *Lancet* 349 (1997): 418; Joelle Gabay, "Ubiquitous Natural Antibiotics," *Science* 64 (1994): 373; Hans Bomans, "Peptide Antibiotics and Their Role in Innate Immunity," *Annual Review of Immunology* 13 (1995): 61; and Tomas Ganz, "Defensins and Host Defense," *Science* 286 (1999): 420.

150–52   Good descriptions of the recent discoveries about allergy, asthma, and autoimmune disorders can be found in Stephen Holgate, "The Epidemic of Allergy and Asthma," *Nature* 402 (suppl.) (1999): B2; William Cookson, "The Alliance of Genes and Environment in Asthma and Allergy," *Nature* 402 (suppl.) (1999): B5; and Lee Nelson et al., "Microchimerism and HLA-Compatible Relationships of Pregnancy in Sclerodoma," *Lancet* 352 (1998): 559.

152–53   Discussion of the interaction between innate and acquired immunity appears in Charles Janeway, Jr., "Approaching the Asymptote? Evolution and Revolution in Immunology," *Cold Spring Harbor Symposia on Quantitative Biology* 54 (1989): 1; Ruslan Medzhitov and Charles Janeway, Jr., "Innate Immunity:

Impact on the Adaptive Immune Response," *Current Opinion in Immunology* 9, no. 1 (1997): 4, "Innate Immunity: The Virtues of a Nonclonal System of Recognition," *Cell* 91 (1997): 295, and "Self-Defense: The Fruit Fly Style," *Proceedings of the National Academy of Sciences, USA* 95 (1998): 429; Jules Hoffman et al., "Phylogenetic Perspective in Innate Immunity," *Science* 284 (1999): 1313; Douglas Fearon, "Seeking Wisdom in Innate Immunity," *Nature* 388 (1997): 323; Ruslan Medzhitov et al., "A Human Homologue of the *Drosophila* Toll Protein Signals Activation of Adaptive Immunity," *Nature* 388 (1997): 394; Dan Hultmark, "Ancient Relationships," *Nature* 367 (1994): 116; and Douglas Fearon and Richard Locksley, "The Instructive Role of Innate Immunity in the Acquired Immune Response," *Science* 272 (1996): 50.

153–54   For accounts of research on the evolution of the immune system, see Samuel Schluter et al., "Molecular Origins and Evolution of Immunoglobulin Heavy-Chain Genes of Jawed Vertebrates," *Immunology Today* 18, no. 11 (1997): 543; Gary Litman, "Sharks and the Origins of Vertebrate Immunity," *Scientific American*, November 1996, p. 67; and Simona Bartl et al., "Molecular Evolution of the Vertebrate Immune System," *Proceedings of the National Academy of Sciences, USA* 91 (1994): 10769. I found information on the origins of RAG genes in Ronald Plasterk, "Ragtime Jumping," *Nature* 394 (1998): 718; and Alka Agrawal et al., "Transposition Mediated by RAG1 and RAG2 and its implications for the evolution of the immune system," *Nature* 394 (1998): 744.

14. MY SALMONELLA

155–58   Descriptions of the workings of cytomegalovirus and other pathogens can be found in Klas Kärre and Raymond Welsh, "Viral Decoy Vetoes Killer Cell," *Nature* 386 (1997): 446; H. E. Farrell et al., "Inhibition of Natural Killer Cells by a Cytomegalovirus MHC Class I Homologue in Vivo," *Nature* 386 (1997): 510; High Rayburn et al., "The Class I MHC Homologue of Human Cytomegalovirus Inhibits Attack by Natural Killer Cells,"

Nature 386 (1997): 514; L. Laurie Scott et al., "Perinatal Herpes-virus Infections," *Infections in Obstetrics*, 11, no. 1 (1997): 27; Brett Finlay and Pascale Cossart, "Exploitation of Mammalian Host Cell Functions by Bacterial Pathogens," *Science* 276 (1997): 718; Julie A. Theriot, "The Cell Biology of Infection by Intra-cellular Bacterial Pathogens," *Annual Review of Cell and Developmental Biology* 11 (1995): 213; and Jorge Galán and James Bliska, "Cross-Talk Between Bacterial Pathogens and Their Host Cells," *Annual Review of Cell and Developmental Biology* 12 (1996): 221.

159   Accounts of the symbiotic relationship between *Vibrio fischeri* and *Euprymna scolopes* can be found in Edward Ruby and Margaret McFall-Ngai, "A Squid That Glows in the Night: Development of an Animal-Bacterial Mutualism," *Journal of Bacteriology* 174 (1992): 4865; Mary Montgomery and Margaret McFall-Ngai, "Bacterial Symbionts Induce Host Organ Morphogenesis During Postembryonic Development of the Squid *Euprymna scolopes*," *Development* 120, no. 7 (1994): 1719; and Laurence Lamarcq and Margaret McFall-Ngai, "Induction of a Gradual, Reversible Morphogenesis of Its Host's Epithelial Brush Border by *Vibrio fischeri*," *Infection and Immunity*, February 1998, p. 777.

159–61   I found excellent information on symbiosis in Jan Sapp, *Evolution by Association: A History of Symbiosis* (Oxford University Press, 1994); Lynn Margulis and Rene Fester, eds., *Symbiosis as a Source of Evolutionary Innovation* (MIT Press, 1991); David Read, "Mycorrhizal Fungi: The Ties That Bind," *Nature* 388 (1997): 517; Philip DeVries, "Singing Caterpillars, Ants and Symbiosis," *Scientific American*, October 1992, p. 76; Ted Schultz, "Ants, Plants, and Antibiotics," *Nature* 398 (1999): 747; and Cameron Currie et al., "Fungus-Growing Ants Use Antibiotic-Producing Bacteria to Control Garden Parasites," *Nature* 398 (1999): 701.

161–64   For more on the microbial ecology of the human body and the effects of antimicrobial products, see Dwayne Savage, "Microbial Ecology of the Gastrointestinal Tract," *Annual Review of*

Microbiology 31 (1977): 107; D. van der Waaij, "The Ecology of the Human Intestine and Its Consequences for Overgrowth by Pathogens Such as *Clostridium difficile,*" *Annual Review of Microbiology* 43 (1989): 69; John Postgate, *Microbes and Man,* 3rd ed. (Cambridge University Press, 1992); Abigail Salyers, *Antibiotic Resistance Gene Transfer in the Mammalian Intestinal Tract* (R. G. Landes, 1995); and Katrin Pütsep et al., "Antibacterial Peptide from *H. pylori,*" *Nature* 398 (1999): 671.

164–65   I found useful material on bacterial communication inside and outside the body in Dale Kaiser and Richard Losick, "How and Why Bacteria Talk to Each Other," *Cell* 73 (1993): 873; Richard Losick and Dale Kaiser, "Why and How Bacteria Communicate," *Scientific American,* February 1997, p. 68; Jeffrey Gordon et al., "Epithelial Cell Growth and Differentiation. III: Promoting Diversity in the Intestine: Conversations Between the Microflora, Epithelium, and Diffuse GALT," *American Journal of Physiology* 273 (1997): G565; and Lynn Bry et al., "A Model of Host-Microbial Interactions in an Open Mammalian Ecosystem," *Science* 273 (1996): 1380.

15. ROOT TALK

167–69   For information on migraine, I consulted Oliver Sacks, *Migraine* (Vintage, 1999), a wonderful source for anyone interested in migraine or in the relationship between body and mind.

169–70   Descriptions of medicinal plant compounds can be found in Michael Balick and Paul Alan Cox, *Plants, People, and Culture: The Science of Ethnobotany* (Scientific American Library, 1996); A. Douglas Kinghorn and Manuel Balandrin, eds., *Human Medicinal Agents from Plants* (American Chemical Society, 1993); Gordon Cragg et al., "Natural Products in Drug Discovery and Development," *Journal of Natural Products* 60 (1997): 52; Manuel Balandrin et al., "Natural Plant Chemicals: Sources of Industrial and Medicinal Materials," *Science* 228 (1985): 1154; Timothy Johns, *With Bitter Herbs They Shall Eat It: Chemical Ecology and the Origins of Human Diet and Medicine* (University of Arizona Press, 1991); Siri von Reis Altschul, "Exploring the Herbarium," *Scientific American,* May 1977, p. 96; William Agosta, *Bombardier Beetles and Fever Trees* (Addison Wesley, 1996); R. H. Whittaker and P. P. Feeny, "Allelochemics: Chemical Interactions Between Species," *Science* 171 (1971): 757; and May Berenbaum, "The Chemistry of Defense: Theory and Practice," *Proceedings of the National Academy of Sciences, USA* 92 (1995): 2.

169–70   My sources on the burial at Shanidar are Arlette Leroi-Gourhan, "The Flowers Found in Shanidar IV, a Neanderthal Burial in Iraq," *Science* 190 (1975): 562; and Ralph Solecki, "Shanidar IV, a Neanderthal Flower Burial in Northern Iraq," *Science* 190 (1975): 880.

170–71   For a discussion of animal uses of medicinal plants, see Eloy Rodriguez and Richard Wrangham, "Zoopharmacognosy: The Use of Medicinal Plants in Animals," in K. P. Downum et al., *Recent Advances in Phytochemistry* (Plenum Press, 1993); Richard Wrangham and T. Nishida, "*Aspilia* spp. Leaves: A Puzzle in the Feeding Behavior of Wild Chimpanzees," *Primates* 24 (1983): 276; Eloy Rodriguez et al., "Thiarubrine A, a Bioactive Constituent of *Aspilia* (Asteraceae) Consumed by Wild Chimpanzees," *Experientia* 41 (1985): 419; Jared Diamond, "Dirty Eating for Healthy Living," *Nature* 400 (1999): 120; and J. Chen et al., "Termites Fumigate Their Nests with Naphthalene," *Nature* 392 (1998): 558.

172–74   I found descriptions of the interaction between plant molecules and human cells in Randolph Nesse and Kent Berridge, "Psychoactive Drug Use in Evolutionary Perspective," *Science* 278 (1997): 63; Leslie Iversen, "How Does Morphine Work?" *Nature* 383 (1996): 759, and *The Science of Marijuana* (Oxford University Press, 2000); E. J. Nestler, "Molecular Mechanisms of Drug Addiction," *Journal of NeuroScience* 12 (1992): 2439; David Julius, "Another Opiate for the Masses?" *Nature* 386 (1997): 442; R. J. Walker, H. L. Brooks, and L. Holden-Dye, "Evolution and Overview of Classical Transmitter Molecules and Their Receptors," *Parasitology* 113 (1996): S3; Trevor Stone and Gail Darlington, *Pills, Potions, and Poisons: How Drugs Work* (Oxford University

Press, 2000); V. Di Marzo, "Formation and Inactivation of Endogenous Cannabinoid Anandamide in Central Neurons," *Nature* 372 (1994): 686; A. Calignano et al., "Control of Pain Initiation by Endogenous Cannabinoids," *Nature* 394 (1998): 277.

173   I refer to Colleen McClung and Jay Hirsh, "Stereotypic Behavioral Responses to Free-Base Cocaine and the Development of Behavioral Sensitization in *Drosophila*," *Current Biology* 8, no. 2 (1998): 109.

174   The study on glutamate is Hon-Ming Lam et al., "Glutamate Receptor Genes in Plants," *Nature* 396 (1998): 125.

175   Descriptions of the search for and discovery of natural medicinal products are found in Gordon Cragg et al., "Drug Discovery and Development at the National Cancer Institute: Potential for New Pharmaceutical Crops," in J. Janick, ed., *Progress in New Crops* (ASHS Press, 1996); Charles Marwick, "Nature's Agents Help Heal Humans," *Journal of the American Medical Association* 279, no. 21 (1998): 1679; Lynn Caporale, "Chemical Ecology: A View from the Pharmaceutical Industry," *Proceedings of the National Academy of Sciences, USA* 92 (1995): 75; and Thomas Eisner, "Prospecting for Nature's Chemical Riches," *Issues in Science and Technology* 6, no. 2 (1998): 31.

16. THE LOOP OF TIME

179–80   For information on comet Hale-Bopp, I consulted Dale Cruikshank, "Stardust Memories," *Science* 275 (1997): 1895; H. A. Weaver, "The Activity and Size of the Nucleus of Comet Hale-Bopp," *Science* 275 (1997): 1900; and Robert Irion, "Rare Gas Pinpoints Hale-Bopp's Cradle," *Science* 288 (2000): 2123.

182–84   A wonderful book on concepts of time in different cultures and throughout history is G. J. Whitrow, *Time in History* (Oxford University Press, 1988).

182–89   To sort out the details of our biological clocks and to trace the history of their discovery, I depended on several articles, among them Charles Czeisler, "Stability, Precision, and Near-24-Hour Period of the Human Circadian Pacemaker," *Science* 284 (1999): 2177; Benjamin Aronson et al., "Circadian Rhythms," *Brain*

Research Reviews 18 (1993): 315; Gianluca Tosini and Michael Menaker, "Circadian Rhythms in Cultured Mammalian Retina," *Science* 272 (1996): 419; and Fred Turek, "Biological Rhythms in Reproductive Processes," *Hormone Research* 37 (suppl. 3) (1992): 93.

183–84   I refer to Gene Block et al., "Biological Clocks in the Retina: Cellular Mechanisms of Biological Timekeeping," *International Review of Cytology* 146 (1993): 83; and Stephan Michel, "Circadian Rhythm in Membrane Conductance Expressed in Isolated Neurons," *Science* 259 (1993): 239.

185–88   My story on the genetics and molecular biology of circadian clocks comes from dozens of references, including Martha Hotz Vitaterna et al., "Mutagenesis and Mapping of a Mouse Gene, Clock, Essential for Circadian Behavior," *Science* 264 (1994): 719; David King et al., "Positional Cloning of the Mouse Circadian Clock Gene," *Cell* 89 (1997): 641; Michael Young, "Marking Time for a Kingdom," *Science* 288 (2000): 451; Phillip Lowrey et al., "Positional Syntenic Cloning and Functional Characterization of the Mammalian Circadian Mutation *tau*," *Science* 288 (2000): 483; Jeffrey Hall and Michael Rosbash, "Oscillating Molecules and How They Move Circadian Clocks Across Evolutionary Boundaries," *Proceedings of the National Academy of Sciences, USA* 90 (1993): 5382; Jay Dunlap, "An End in the Beginning," *Science* 280 (1998): 1548; Nicholas Gekakis et al., "Role of the CLOCK Protein in the Mammalian Circadian Mechanism," *Science* 280 (1998): 1564; Thomas Darlington et al., "Closing the Circadian Loop," *Science* 280 (1998): 1599; Ueil Schibler, "New Cogwheels in the Clockworks," *Nature* 393 (1998): 620; Steve Kay, "PAS, Present and Future: Clues to the Origins of Circadian Clocks," *Science* 276 (1997): 753; Paolo Sassone-Corsi, "Molecular Clocks: Mastering Time by Gene Regulation," *Nature* 392 (1998): 871; David Whitmore et al., "Light Acts Directly on Organs and Cells in Culture to Set the Vertebrate Circadian Clock," *Nature* 404 (2000): 87; and Lauren Shearman et al., "Interacting Molecular Loops in the Mammalian Circadian Clock," *Science* 288 (2000): 1013.

186   The work on *period* was conducted by Seymour Benzer. See his paper "Genetic Dissection of Behavior," *Scientific American* 229 (1973): 24–37.

186   I refer to Martin Ralph and Michael Menaker, "A Mutation of the Circadian System in Golden Hamsters," *Science* 241 (1988): 1225; and Martin Ralph et al., "Transplanted Suprachiasmatic Nucleus Determines Circadian Period," *Science* 247 (1990): 975. For Takahashi's story, I drew on Joseph Takahashi and Michelle Hoffman, "Molecular Biological Clocks," *American Scientist* 83 (1995): 158; and Joseph Takahashi et al., "Forward and Reverse Genetic Approaches to Behavior in the Mouse, *Science* 264 (1994): 1724.

187   I refer to Andrew Millar et al., "Circadian Clock Mutants in *Arabidopsis* Identified by Luciferase Imaging," *Science* 267 (1995): 1161; and Jeffrey Plautz et al., "Independent Photoreceptive Circadian Clocks Throughout *Drosophila*," *Science* 278 (1997): 1632.

17. OF AGE

Among the general sources on aging I consulted were Leonard Hayflick, "New Approaches to Old Age," *Nature* 403 (2000): 365; Leonard Guarente, "What Makes Us Tick?" *Science* 275 (1997): 943; John Medina, *The Clock of Ages* (Cambridge University Press (1996); Steven Austad, *Why We Age* (Wiley, 1997); Robert Ricklefs and Caleb Finch, *Aging: A Natural History* (Scientific American Library, 1995); Simone de Beauvoir, *The Coming of Age,* trans. Patrick O'Brian (Putnam, 1972); and Rich Miller and Steve Austad, "Large Animals in the Fast Lane," *Science* 285 (1999): 199. For an up-to-date discussion of the genetic and biological processes underlying aging, see the following review articles in *Nature* 408 (2000). Thomas Kirkwood and Steven Austad, "Why Do We Age?" 233; Toren Finkel and Nikki Holbrook, "Oxidants, Oxidative Stress and the Biology of Aging," 239; Leonard Guarente and Cynthia Kenyon, "Genetic Pathways that Regulate Ageing in Model Organisms," 225; George Martin and Junko Oshima, "Lessons from Human Progeroid Syndromes," 263; and Leonard Hayflick, "The Future of Ageing," 267.

194–95   Information on Alzheimer's and the ApoE genes can be found in Mohammad Kamboh, "Apolipoprotein E Polymorphism and Susceptibility to Alzheimer's Disease," *Human Biology* 67 (1995): 195; Mohammad Kamboh et al., "A Novel Mutation in the Apolipoprotein E Gene (APOE*4 Pittsburgh) Is Associated with the Risk of Late-Onset Alzheimer's Disease," *NeuroScience Letters* 263, no. 2-3 (1999): 129; D. Blacker et al., "Alpha-2-macroglobulin Is Genetically Associated with Alzheimer's Disease," *Nature Genetics* 19 (1998): 357; Jacob Raber et al., "Apolipoprotein E and Cognitive Performance," *Nature* 404 (2000): 352; and Bruce Yankner, "A Century of Cognitive Decline: If We Live Long Enough, Will We All Become Demented?" *Nature* 404 (2000): 125.

195–96   I refer to David Snowdon et al., "Cognitive Ability in Early Life and Cognitive Function and Alzheimer's Disease in Late Life: Findings from the Nun Study," *Journal of the American Medical Association* 275 (1996): 528.

197–98   Caleb Finch discusses the rate-of-living theory and lifespan in *Aging: A Natural History*. For information on progeria, I consulted W. Ted Brown, "Progeria: A Human-Disease Model of Accelerated Aging," *American Journal of Clinical Nutrition* 55 (1992): 1222S. Leonard Hayflick's work is described in his article "Cell Biology of Aging," *BioScience* 25, no. 10 (1975): 629.

198–99   Discussions of telomeres and their role in aging can be found in Elizabeth Blackburn, "Broken Chromosomes and Telomeres," in Fedoroff and Botstein, *Dynamic Genome;* Victoria Lundblad, "Telomeres Keep on Rappin'," *Science* 288 (2000): 2141; Titia de Lange, "Telomeres and Senescence: Ending the Debate," *Science* 279 (1998): 334; Carol Greider and Elizabeth Blackburn, "Telomeres, Telomerase and Cancer," *Scientific American,* February 1996, p. 92; Virginia Zakian, "Telomeres: Beginning to Understand the End," *Science* 270 (1995): 1601; and Michael Fossel, "Telomerase and the Aging Cell," *Journal of the American Medical Association* 229, no. 21 (1998): 1732.

200–201   For more on cancer and the p53 gene, see Robert Weinberg, *One Renegade Cell* (Weidenfeld and Nicolson, 1998); M. Hollstein et al., "p53 Mutations in Human Cancers," *Science* 253 (1991):

49; H. Symonds et al., "p53-dependent Apoptosis Suppresses Tumor Growth and Progression in Vivo," *Cell* 78 (1994): 703; and Ashok Venkitaraman, "Breast Cancer Genes and DNA Repair," *Science* 286 (1999): 1100.

201–2   For studies on genes and aging, I referred to Caleb Finch and Rudolph Tanzi, "Genetics of Aging," *Science* 278 (1997): 407; Koutarou Kimura et al., "*daf-2,* an Insulin Receptor-like Gene That Regulates Longevity and Diapause in *Caenorhabditis elegans,*" *Science* 277 (1997): 942; Bernard Lakowski and Siegfried Hekimi, "Determination of Life-Span in *Caenorhabditis elegans* by Four Clock Genes," *Science* 272 (1996): 1010; Yi-Jyun Lih et al., "Extended Life-Span and Stress Resistance in the *Drosophila* Mutant *methuselah,*" *Science* 282 (1998): 943; Maokoto Kuro-o et al., "Mutation of the Mouse *klotho* Gene Leads to a Syndrome Resembling Aging," *Nature* 390 (1997): 45; and Leonard Guarente, "Mutant Mice Live Longer," *Nature* 402 (1999): 243.

202   I refer to Danith Ly et al., "Mitotic Misregulation and Human Aging," *Science* 287 (2000): 2486.

202–3   I found discussion of the role in aging of free radicals and mitochondria in Bruce Ames et al., "Oxidants, Antioxidants, and the Degenerative Diseases of Aging," *Proceedings of the National Academy of Sciences, USA* 90 (1993): 7915; and Yuichi Michikawa et al., "Aging-Dependent Large Accumulation of Point Mutations in the Human mtDNA Control Region for Replication," *Science* 286 (1999): 774.

204–5   For Williams's theory, I referred to George Williams and Randolph Nesse, "The Dawn of Darwinian Medicine," *Quarterly Review of Biology* 66, no. 1 (1991): 1.

205–6   On past and present efforts to prolong life, see Gairdner B. Moment, "The Ponce de Leon Trail Today," *BioScience* 25, no. 10 (1975): 623; Michael Rose, "Can Human Aging Be Postponed?" *Scientific American,* December 1999, p. 106; A. G. Bodnar et al., "Extension of Life-Span by Introduction of Telomerase into Normal Human Cells, *Science* 279 (1998): 349; David Gems and Richard Weindruch, "Caloric Restriction and Aging," *Scientific American,* January 1996, p. 46; Judith Campisi, "Aging, Chromatin, and Food Restriction — Connecting the Dots," *Science* 289 (2000): 2062; Su Ju Lin et al., "Requirement of NAD and *SIR2* for Lifespan Extension by Calorie Restriction in *Saccharomyces cerevisiae*," *Science* 289 (2000): 2126; and Donald Riddle, "Longevity in *Caenorhabditis elegans* Reduced by Mating but Not Gamete Production," *Nature* 379 (1996): 723.

206   I refer to Daniel Promislow, "Longevity and the Barren Aristocrat," *Nature* 396 (1998): 719.

18. SWEET MYSTERY

207–8   For discussion of archaebacteria and the cradle of life, I referred to Edward DeLong, "Archaeal Means and Extremes," *Science* 280 (1998): 542; Euan Nisbet, "The Realms of Archaean Life," *Nature* 405 (2000): 625; Euan Nisbet and C. M. R. Fowler, "Some Liked It Hot," *Nature* 382 (1996): 404; Govert Schilling, "RNA Study Suggests Cool Cradle of Life," *Science* 283 (1999): 155; and Nicolas Galtier, "A Nonhyperthermophilic Common Ancestor to Extant Life Forms," *Science* 283 (1999): 220.

208–9   I found references on the ubiquity of water in Takeshi Oka, "Water on the Sun: Molecules Everywhere," *Science* 277 (1997): 328; Richard Kerr, "An Ocean Emerges on Europa," *Science* 276 (1997): 355; John Noble Wilford, "Jupiter's Moon Might Be Cradle for New Life," *New York Times*, April 10, 1997, p. A1; and Richard Monastersky, "Reservoir of Water Hides High Above Earth," *Science News* 152 (1997): 117.

210   Theories of our aquatic ancestry are described in Evan S. Connell, *The White Lantern* (North Point Press, 1989), pp. 3–4; and in Elaine Morgan, *The Scars of Evolution: What Our Bodies Tell Us about Human Origins* (Oxford University Press, 1994).

211–12   A lively and engaging account of the chemistry and physics of water can be found in Philip Ball, *Life's Matrix: A Biography of Water* (Farrar, Straus and Giroux, 2000).

212   I refer to Leslie Kuhn et al., "The Interdependence of Protein Surface Topography and Bound Water Molecules Revealed by Surface Accessibility and Fractal Density Measures," *Journal of Molecular Biology* 228 (1992): 13.

212–13   Carl Woese's theory is discussed in his paper "The Universal Ancestor," *Proceedings of the National Academy of Sciences, USA* 95 (1998): 6854.

214   The quotation from Robert Miller comes from his book *The Sea* (Random House, 1966).

216   For a discussion of the genetic differences among humans, see Kelly Owens and Mary-Claire King, "Genomic View of Human History," *Science* 286 (1999): 451. I refer to the study by Samantha Knight, "Subtle Chromosomal Rearrangements in Children with Unexplained Mental Retardation," *Lancet* 354 (1999): 1676; to Barbara McClintock's paper, "The Significance of Responses of the Genome to Challenge," in Fedoroff and Botstein, *Dynamic Genome;* and to Suzanne Rutherford and Susan Lindquist, "Hsp90 as a Capacitor for Morphological Evolution," *Nature* 396 (1998): 336. See also the accompanying commentary, Andrew Cossins, "Cryptic Clues Revealed," *Nature* 396 (1998): 309.

ACKNOWLEDGMENTS

THANKS FIRST to the numerous scientists in the fields of molecular and cell biology, evolutionary biology, immunology, developmental biology, zoology, and other disciplines, whose research and theories I have drawn upon in the writing of this book. I have relied heavily on their original work. I am deeply grateful to all of them, and especially to Richard Dawkins, Stephen Jay Gould, William Hamilton, François Jacob, John Maynard Smith, Lewis Thomas, and George C. Williams.

I am also indebted to the scientists who gave their time to be interviewed, to send references, and to read drafts of the chapters. Thanks especially to Russell Doolittle, who read the book in its entirety and offered encouragement and incisive suggestions, and also to Susan Alberts, Laura Attardi, Gene Block, Sean Burgess, Sean Carroll, Gloria Coruzzi, David Crews, Ann Ferentz, Jeffrey Gordon, Michael Grace, Diane Hoffman-Kim, H. Robert Horvitz, Ruth Hubbard, Charles Janeway, Jr., Anne-Françoise Lamblin, Michael Land, Michael Menaker, Christiane Nüsslein-Volhard, David Pfennig, Ronenn Roubenoff, David Schatz, Allan Spradling, and Michael Zasloff. I thank you all for your generous assistance. Whether or not my book rings true, you have done your best to make it so; any lingering mistakes are my own.

To the Alfred P. Sloan Foundation and to Doron Weber I am grateful for the grant that allowed me to take the time to write and research this book. I owe deep gratitude, as well, to the Bunting Institute of Radcliffe College for a fellowship in creative nonfiction in 1997–1998, which supported my work on the book and put me in the company of an exceptional group of sister fellows, many of whom gave expert advice. Thanks especially to fellow writers and dear friends Mim Nelson and Barbara

Goldoftas, and to the three astonishing young women who served as my research associates through the Radcliffe Research Partnership Program: Jennifer Fu, Jessica Hammer, and Tam Pui-Ying.

I am grateful to Carin Algava, who offered able and cheerful help at an early stage in the project. Thanks also to Maya Pines, now retired from the Howard Hughes Medical Institute, for introducing me to the world of molecular biology, and to Bob Poole at the National Geographic Society for fostering my writing career from its beginnings.

Special thanks to Susan Bacik, whose brilliant and beautiful sculpture gave the book its central metaphor; to Janet Silver, for her superb editorial instincts; and to Melanie Jackson, for representing this project with such abounding grace and flair. Thanks to my daughters, Zoë and Nell, for being their loving, inquisitive selves. To my husband, the writer Karl Ackerman, my love and gratitude for being, as always, funny, insightful, endlessly supportive.